THE
WINTER
BIRDS

M.A.Ogilvie

LONDON
MICHAEL JOSEPH

First published in Great Britain by Michael Joseph Ltd.,
52 Bedford Square, London WC1B 3EF

ISBN 0 7181 1529 5

Designed and produced by Walter Parrish International Limited, London

Designer Stephen Chapman

Set in Imprint 101

Printed and bound in Spain by
Novograph S.A., Madrid
Dep Legal: M 15577/1976

to Carol

Contents

List of species described in detail

Where British and American species names differ, the American is given in square brackets.

Waterbirds

Divers [Loons]
Red-throated Diver [Red-throated Loon]
Black-throated Diver [Arctic Loon]
Great Northern Diver [Common Loon]
White-billed Diver [Yellow-billed Loon]
Swans
Bewick's Swan and Whistling Swan
Whooper Swan
Grey Geese
Bean Goose
Pink-footed Goose
White-fronted Goose
Lesser White-fronted Goose
Greylag Goose
Snow Geese
Lesser Snow Goose and Blue Goose
Greater Snow Goose
Ross's Goose
Emperor Goose
Black Geese
Canada Goose
Barnacle Goose
Brent Goose [Black Brant]
Red-breasted Goose
Dabbling Ducks
Mallard
Pintail
Wigeon
American Wigeon [Baldpate]
Teal
Green-winged Teal
Baikal Teal
Diving Ducks, Sea Ducks, and Sawbills
Scaup [Greater Scaup]
Harlequin Duck
Long-tailed Duck [Old Squaw]
Barrow's Goldeneye
Red-breasted Merganser
Common Scoter [Black Scoter]
Surf Scoter
Velvet Scoter [White-winged Scoter]
Eiders
Common Eider
King Eider
Spectacled Eider
Steller's Eider

Waders or shorebirds

Plovers
Golden Plover
American Golden Plover
Grey Plover [Black-bellied Plover]
Ringed Plover including Semi-palmated Plover
Small Waders
Knot
Sanderling
Semi-palmated Sandpiper
White-rumped Sandpiper
Baird's Sandpiper
Pectoral Sandpiper
Curlew Sandpiper
Purple Sandpiper including Rock Sandpiper
Dunlin [Red-backed Sandpiper]
Stilt Sandpiper
Buff-breasted Sandpiper
Turnstone [Ruddy Turnstone]
Great Knot
Western Sandpiper
Little Stint
Red-necked Stint
Temminck's Stint
Least Sandpiper
Sharp-tailed Sandpiper
Spoon-billed Sandpiper
Medium and large waders
Ruff
Long-billed Dowitcher

Eskimo Curlew
Whimbrel [Hudsonian Curlew]
Hudsonian Godwit
Bar-tailed Godwit
Spotted Redshank
Redshank
Lesser Yellowlegs

Phalaropes
Red-necked Phalarope [Northern Phalarope]
Grey Phalarope [Red Phalarope]

Seabirds

Skuas [Jaegers]
Pomarine Skua [Pomarine Jaeger]
Arctic Skua [Parasitic Jaeger]
Long-tailed Skua [Long-tailed Jaeger]

Large Gulls
Herring Gull
Iceland Gull
Glaucous Gull
Ivory Gull
Great Black-backed Gull

Small Gulls
Sabine's Gull
Ross's Gull
Kittiwake
Common Gull

Terns
Arctic Tern

Auks
Common Guillemot [Common or Pacific Murre]
Brunnich's Guillemot [Thick-billed Murre]
Razorbill
Black Guillemot [Mandt's Guillemot] including Pigeon Guillemot

Auklets and Puffins
Kittlitz's Murrelet
Little Auk [Dovekie]
Crested Auklet
Least Auklet
Parakeet Auklet
Common Puffin
Horned Puffin
Tufted Puffin

Fulmars
Fulmar

Cormorants
Cormorant
Pelagic Cormorant
Shag

Land birds

Raptors
White-tailed Eagle
Rough-legged Buzzard [Rough-legged Hawk]
Gyr Falcon
Peregrine Falcon [Duck Hawk]
Merlin [Pigeon Hawk]

Owls
Snowy Owl
Short-eared Owl

Gamebirds and Cranes
Willow Ptarmigan [Willow Grouse] including Red Grouse
Rock Ptarmigan [Ptarmigan]
Sandhill Crane

Larks, Pipits, and Wagtails
Shore Lark [Horned Lark]
Red-throated Pipit
Water Pipit including Rock Pipit
Yellow Wagtail
Yellow-headed Wagtail [Citrine Wagtail]
White Wagtail including Pied Wagtail
Pechora Pipit
Meadow Pipit

Wheatears and Thrushes
Wheatear
Fieldfare
Redwing

Redpolls
Redpoll
Arctic Redpoll

Sparrows
Savannah Sparrow
White-crowned Sparrow
Tree Sparrow
Harris's Sparrow

Buntings [Longspurs]
Lapland Bunting [Lapland Longspur]
Smith's Longspur
Snow Bunting

Crows
Raven

Preface

I started bird-watching on the coast of Essex in eastern England. Here, beside the estuaries and broad tidal mudflats of that county, many winter days were spent becoming familiar with such species as the Brent Goose, Pintail, the occasional Eider Duck and Scoter, plus innumerable waders. Of less frequent but still regular occurrence were divers, wild swans, Long-tailed Duck, Purple Sandpiper, Short-eared Owl, and Snow Bunting. The most I knew about the nesting habits of such birds was that they all had one thing in common—they bred in the arctic, an impossibly far-off sounding place in those days.

However, I did get to the arctic, spending two months in the summer of 1961 in northeast Greenland. Then as on later trips my reason for going was to study geese, but it also gave me the opportunity to see some familiar species in unfamiliar surroundings, to see birds on their breeding grounds that hitherto I had only known as migrants or winter visitors. It was intensely satisfying to be able to follow them through the previously missing part of their life cycle, to see the nests of Barnacle and Pink-footed Geese, and of Knot and Sanderling, to watch the young ones grow and fledge, ready for their journey south, and to know that many of them would be going to the same country as I was at the end of the short summer.

In that first visit I learnt a little about the arctic life of some of the wintering birds of Britain, but was eager to know very much more. Two things immediately stand out from any study. First the number of species occurring there is comparatively small yet many of them have very wide distributions, being completely or nearly completely circumpolar. Second, only the merest handful of species remains there for the winter, the great majority migrating south. Thus bird-watchers both in North America and in Europe can not only hope to see a large number of arctic-breeding birds either on passage or wintering, but many of the species they see will be the same in both continents. This universality is one of their great attractions; the huge distances they travel is another; while the secrets of their breeding in that (for most people) remote and unattainable land is a third. I hope that in this book I have been able to draw all these main strands together to provide an overall picture of the arctic and its birds, relating them also to their migrations and wintering areas.

There are still a great many gaps in our knowledge, a few of which, mainly concerning geese, I have had a part in trying to fill in on my trips to the arctic. Some species are still very little known, either in distribution or breeding habits, while studies of others have revealed anomalies and contradictions that merely add to the interest. A number of 'rules' that relate to these birds have been put forward. They are in effect statements that formulate what seem to be commonly occurring, natural phenomena, most relating physical variations with latitude. For example the birds tend to be larger, have paler plumage, and have smaller extremities than those further south. But exceptions to these rules occur, some of which cannot readily be explained, or at least in the present state of knowledge. The more one probes the more new questions appear, and the more fascinating it becomes.

This is not simply a seeking after knowledge for its own sake. We will never know everything about the birds of the arctic, but we do need to know as much as possible if we are to protect them against the damage that man is potentially capable of inflicting upon them. There is no case for stopping the exploration and exploitation of the arctic's mineral wealth, but there is every reason for trying to avoid its worst effects. Many of the birds provide food, sport, or sheer pleasure for millions of people living further south. It is their summer home, the whole fragile ecosystem of the arctic, that must be safeguarded.

M. A. Ogilvie

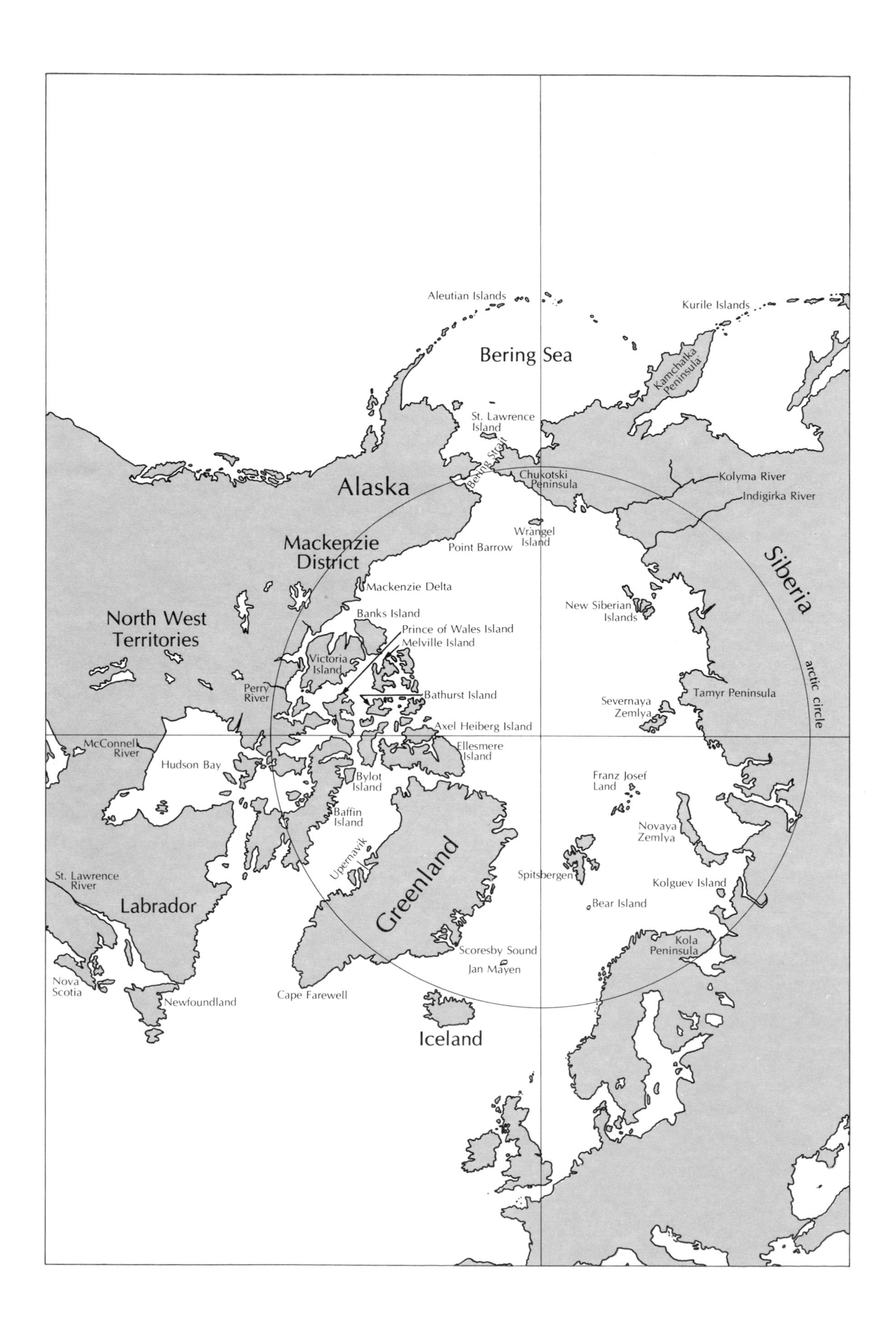

Aleutian Islands
Kurile Islands
Bering Sea
Kamchatka Peninsula
St. Lawrence Island
Bering Strait
Chukotski Peninsula
Kolyma River
Indigirka River
Alaska
Wrangel Island
Point Barrow
Mackenzie District
Siberia
Mackenzie Delta
New Siberian Islands
Banks Island
North West Territories
Prince of Wales Island
Melville Island
Victoria Island
arctic circle
Perry River
Bathurst Island
Tamyr Peninsula
Severnaya Zemlya
Axel Heiberg Island
McConnell River
Ellesmere Island
Hudson Bay
Bylot Island
Franz Josef Land
Baffin Island
Novaya Zemlya
Upernavik
Greenland
Spitsbergen
St. Lawrence River
Kolguev Island
Labrador
Bear Island
Kola Peninsula
Scoresby Sound
Jan Mayen
Nova Scotia
Cape Farewell
Newfoundland
Iceland

1 The nature and extent of the arctic

Most people have some idea of what is meant by the arctic—a land of snow and ice, where the sun shines continuously in the summer and not at all in the winter, and where it can be bitterly cold, so cold that the sea freezes. All of these things are true in part but none fully describes the region or enables its geographical limits to be defined.

It may in fact come as a surprise to learn that it has no single agreed definition. Those that do exist have been made to serve some specialist purpose and are therefore not suitable for universal application. The main features used in such attempts are astronomical, physical, meteorological, and biological. In a book on birds the biological aspects will naturally be used as much as possible, but first it is necessary to explain why some of the other definitions must be discarded.

One is taught in school that the Arctic Circle encompasses the globe at a latitude of 66°33′ North, and that it forms the southern boundary of the area within which that intriguing phenomenon, the midnight sun, shines in summer; conversely the days in winter are all dark. At a locality situated precisely on the Arctic Circle there is just one night at mid-summer when the sun shines all the time, and one day in the winter when it does not rise at all. To the north, the summer period of continuous daylight and the winter period of darkness increase, until at 78°N, or roughly halfway from the Arctic Circle to the North Pole, there is about three months of continuous sun (clouds permitting of course) in summer, and a corresponding period of darkness in winter. At the North Pole itself there is a theoretical six-month day followed by a six-month night, though in actual fact refraction of the sun's rays by the earth's atmosphere extends the 'day' by about a fortnight, while there is a further twilight period of five to six weeks at the start and finish of the 'night'.

Some of the geographical features of the far north.

The Arctic Circle is thus a purely astronomical boundary and although lands to the north might reasonably be called arctic, a glance at the map (p. 14) shows that this is a very imprecise definition. The circle cuts right across Greenland, leaving a large area of that country's icecap to the south, and across northern Canada, again to the north of much land that cannot really be considered as anything but arctic in nature. On the other hand it includes within its sweep extensive areas of northern Scandinavia where, thanks to the warming influence of the Gulf Stream, the climate, fauna, and flora are all closer to those in the temperate zone.

Two physical features that are often thought of as synonymous with the arctic are the phenomena of permanently frozen ground—the permafrost—and its marine corollary—the existence of frozen sea, at least in winter. Both occur widely in northern latitudes but neither stand up to close scrutiny as reasonable definitions that put fairly precise limits on the region's southern extent. Permafrost occurs as a result of cumulative cold temperatures. Its southern boundary coincides fairly closely with the isotherm (a line on a map joining places of equal temperature) for a mean annual temperature of −5°C (23°F).

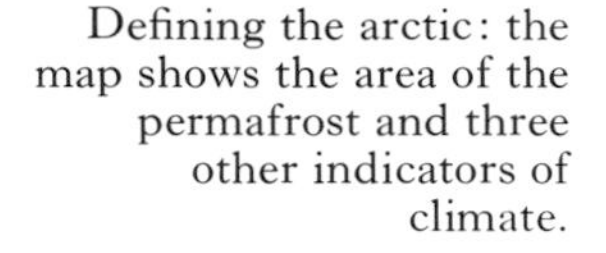

Defining the arctic: the map shows the area of the permafrost and three other indicators of climate.

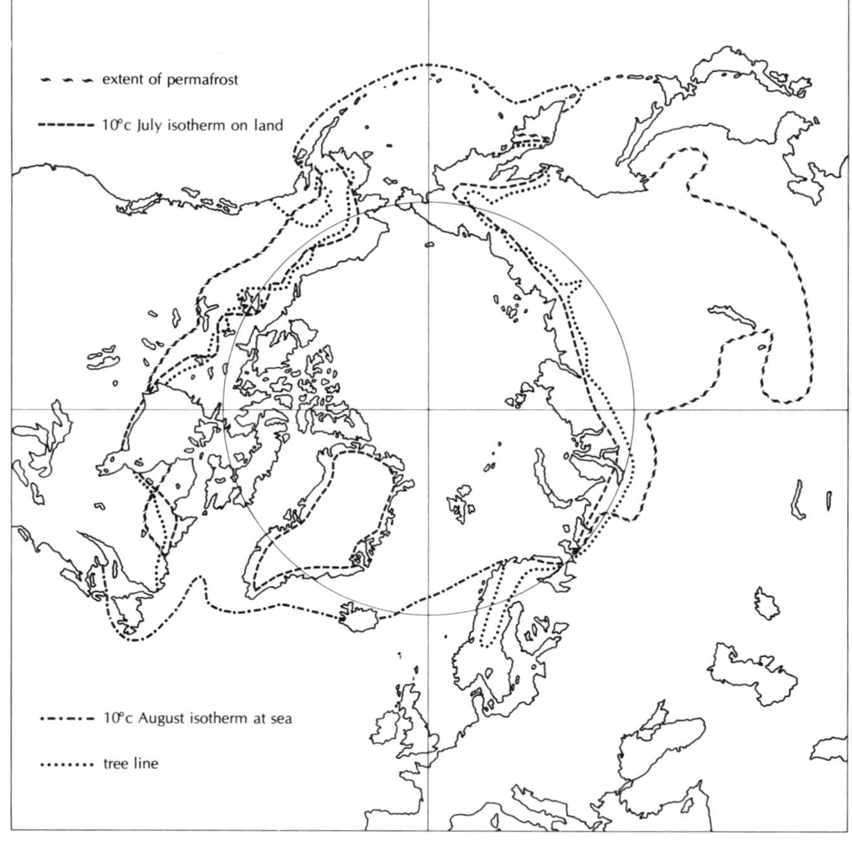

Although such a temperature occurs over much of the arctic, the cooling effect felt in the centres of large continental masses produces such cold winter temperatures that the isotherm is shifted south almost to 50°N in both Asia and North America, in other words south of the latitude of London or New York. Enormous areas of Canada and Siberia have permafrost below the ground but also have sufficiently warm summers for the land to be covered in forests or even, in some parts, to be cultivated and an annual harvest of grain taken.

Freezing of the sea occurs in winter in northern latitudes but the ice so formed may be drifted far to the south by ocean currents or, to a lesser degree, by the wind. Where the ice meets warm water and melts depends on the relative strengths of cold and warm currents. Thus its final distribution depends on sea and air temperatures in regions far removed from where it was first formed. And of course the occurrence of ice formation or even its extent is of no use across the land masses of Eurasia and North America, and similarly the permafrost boundary is no use at sea.

Cold temperatures too are widely associated in people's minds with this region but although, as will be seen, temperature is of vital importance in sustaining life, single measurements of extreme cold temperature are not particularly meaningful. The coldest places on earth are not especially far north. Localities in Siberia where winter temperatures of −50°C (−58°F) have been recorded are covered in trees, and the coldest temperature recorded in the far north is nothing like as low as this.

It is best, therefore, to ignore extremes when considering temperature and to look at means recorded over a period. Most forms of animal and plant life can survive occasional extreme temperatures, whether hot or cold, but are much more affected by the average temperature that they experience over a long period. Isotherms drawn for mean monthly temperatures have been found to have definite relationships with various biological occurrences. Some forms of life will only flourish when the mean temperature in the coldest part of the year does not fall below a certain level, while other plants or animals require a particular minimum amount of warmth in summer before they can reproduce. The most relevant isotherm in this context is that for a mean temperature in the warmest month of the year (nearly always July) of 10°C (50°F). The relative temperature that this represents can be gauged by comparing it with London where a mean of 10°C occurs in the months of November and February, with only December and January colder, and with New York, where the months October to March are colder than this, but the remainder of the year warmer, often considerably so.

It can be seen from the map that the 10°C July isotherm dips well to the south of the Arctic Circle across the landmasses, though not nearly so far as the permafrost limit, but rises to the north of it in areas such as northern Scandinavia. Also shown is the northern limit of tree growth—it is obvious that the agree-

An aerial view of the patterned ground associated with permafrost. These polygons are produced by alternate freezing and thawing over many years.

ment between the two lines is very close indeed, and this is one of the strongest biological reasons for using this particular isotherm to define the southern limit of the arctic. There are, of course, some deviations (it would be remarkable if there were not) but they are slight compared with the overall agreement. Trees are conspicuous, readily identifiable, and of great economic importance to man. Their use as a biological indicator of different geographical regions is therefore both appropriate and convenient.

The tree line, the northern limit of growth, is not of course a precise, straight line. One does not step directly from forest onto tundra. But over a distance of some miles, in some areas only a few, in others as much as 60 or 70, there is a transition zone between the well-wooded parts and the tundra. In this zone the trees gradually become more scattered and in particular more stunted in growth until they finally peter out altogether. Various local factors affect this boundary as indeed they affect the temperature. On flat windswept land where the prevailing winds come from a colder region the line will bend away from this direction. In rolling country, trees will extend further to the north in the valleys than they will on the hills. But allowing for all these minor variations the tree line forms the single most obvious boundary of the arctic zone.

Shattered ice blocks from a glacier in an arctic fjord.

The subarctic muskeg of stunted spruce trees interspersed with patches of tundra and bog provides breeding places for small landbirds and waders.

Having settled on a boundary on land it is still necessary to determine a boundary at sea. This is less easy to establish because, as already mentioned, the sea is affected by currents and wind which produce considerable distortions in any dividing line between the arctic sea and more temperate areas. The distribution of certain marine organisms has been found to follow sea temperature quite closely, again using mean rather than absolute temperatures. The same isotherm as on land, that of 10°C (50°F) for the warmest month of the year (August rather than July because of the longer time it takes for the sea to warm up) coincides fairly precisely with the southern limit of distribution of a number of sea animals (see map). Only the surface temperature of the sea is considered here, as deeper down there are a number of layers of water having both higher and lower temperatures than the surface.

The arctic on land and at sea has now been given a definition, but before some of the physical and biological features that arise from the definition are examined, there is an important subdivision that must be mentioned. This is the separation into the high arctic zone and the low arctic zone (see p. 18). The zones have some distinct differences in their fauna and flora even though there is also considerable overlap between them. The climate of the high arctic is much more severe than in the low arctic, and once again the best boundary line is a mean temperature isotherm, this time of 5°C (41°F) for the warmest month of the year (July). The number of animals and plants that can survive in the colder northern region is strictly limited. Indeed on the coast the high arctic is characterised by an almost complete absence of life in the inter-tidal zone of the shore. On land the number of plant species falls off sharply and growth of scrub ceases entirely. Similarly the number and variety of birds drop. The division between the two zones is well demarcated in some areas, as when moving north along the coast of Greenland, but is less obvious on the large land areas of Siberia or Canada.

The most important physical features of the arctic are the relatively low temperatures, especially in summer, the low rainfall or

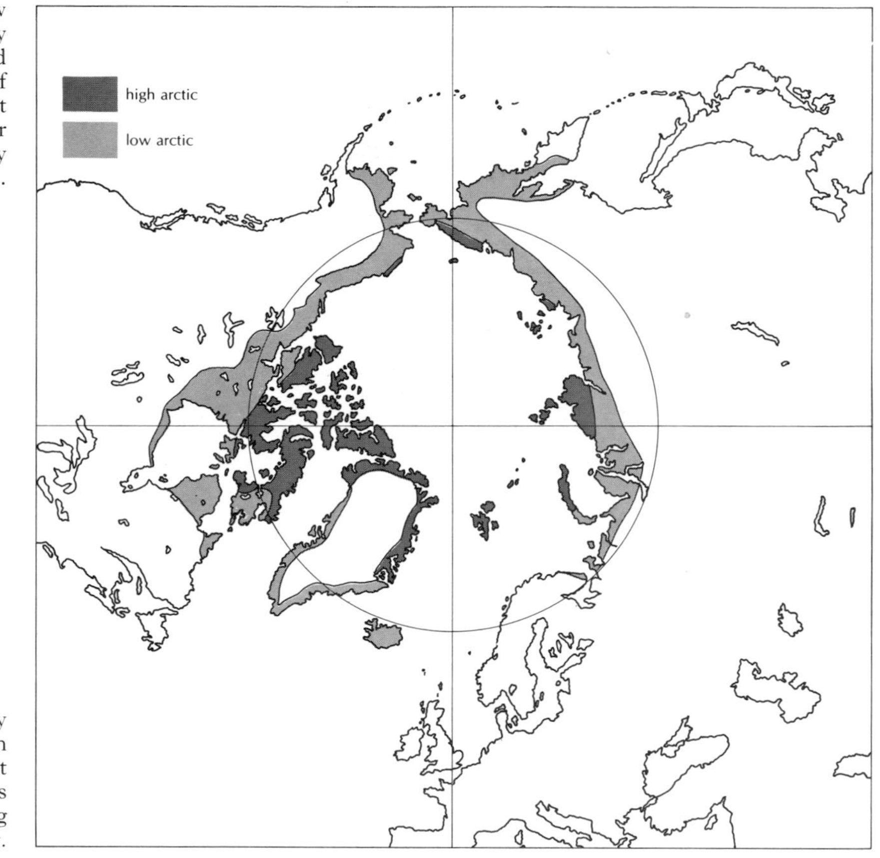

High arctic and low arctic, distinguished by differences in flora and fauna. The number of animals and plants that can survive in the colder northern region is strictly limited.

Opposite: Approximately half the bird species in the arctic are dependent upon wetlands like this tundra lake and adjoining bog.

snowfall, and the permafrost. Each of these has its own marked effect on the environment, and also acts with the others to shape the landscape and dictate what can grow or live there. Although colder temperature extremes can occur elsewhere, the arctic averages are quite low—the mean monthly temperature is below freezing for about seven months of the year in the low arctic and for nine months in the high.

The corollary of the long cold winter is the brief cool summer. Spring and autumn are not really sufficiently distinct to be treated as separate seasons. Although the sun may be shining more or less continuously it has little warmth in it, for the maximum angle that it reaches above the horizon even at mid-summer is still relatively low compared with the angle at lower latitudes. For example at 78°N the highest the sun reaches in the sky is an angle of 24°. This means that it is shining through a fairly thick layer of the earth's atmosphere which absorbs a significant proportion of its warming power. The amount of light is sufficient to provide a long growing season for the plants, and plenty of daylight hours for the birds seeking food for their young, but the lack of heat in the sun has an inhibiting effect on growth of all kinds. In summer in the high arctic the maximum air temperature only rarely reaches 10°C to 12°C (50°F to 55°F) and freezing conditions may occur at almost any time, even in the 'warmest' month. In mountainous regions the

sun often 'sets' behind high ground even though it is technically always above the true horizon, and the drop in temperature when this happens is quite noticeable. Even in the low arctic only the months of July and August are virtually free of frosts.

The additional cooling effect of the wind must not be overlooked. The incidence of gale force or strong winds is very high and even a sunny day can be made to feel cold when strong winds are blowing off a frozen sea or an icecap. The wind has a powerful effect on plants, shrivelling them as much as frost can, and inhibiting upward growth. It also governs the incidence of snow-cover during the winter, which has an important bearing on plant growth.

The concept of a snow-covered land in winter suggests heavy snowfalls, but in fact the total precipitation is surprisingly slight even allowing for one inch of rain equalling nearly one foot of snow. Over much of the arctic it is under 20 inches rainfall equivalent, and only a very few inches in some areas. For example in the northern islands of the Canadian arctic archipelago the average annual precipitation is less than 6 inches, while in northern Greenland it is even lower. This compares with a range of from 20 inches in eastern England to several times that on the west coast, or the average annual precipitation on the eastern seaboard of North America of about 50 inches. In warmer parts of the globe an annual rainfall of less than about

12 inches usually leads to desert conditions, while to the very small amounts of rain must be added the effects of the high rate of evaporation caused by the strong sun. Parts of the high arctic are little better than deserts but one important factor enables plants to make better use of the very low precipitation than they would be able to elsewhere—the factor of the permafrost.

The permafrost layer starts just below the ground surface. In winter all the ground is frozen, but when spring comes a thin surface layer thaws out. In the high arctic this layer may be only a few inches in depth, extending to a foot or a foot and a half in some areas, while in the low arctic the unfrozen layer may extend as deep as three or four feet. The frozen layer acts as a total barrier to the roots of plants, so that they are forced to be shallow rooted, but it also, and more importantly, prevents the drainage of the land. The water cannot sink through the soil as it would normally, so it is trapped near or even on the surface. This accounts for the very extensive bogs that occur on the flat tundra. When walking over them one's feet sink in several inches before reaching the hard frozen layer. Such progression is very tiring indeed, though at least there is no danger of plunging much deeper into some unseen hole as is possible in the more treacherous bogs of the temperate lands. Because the water cannot drain away it is more or less constantly available to the plants on the tundra. They are virtually growing on the surface of a shallow reservoir which holds for them such rain and meltwater as there is. Without this reservoir effect the arctic would be really barren, not that it is spectacularly verdant anywhere.

Most of the precipitation falls as snow during the winter, and where it lies it acts as an insulating cover to the ground and plants beneath. The wind governs the incidence of snow-cover by blowing some areas free and covering others with deep drifts. Where plants are left exposed through the winter they are very liable to be killed off by the frosts, but even an inch or two of snow prevents the temperature dropping anything like as low as it would on bare ground. Once the snow surface has been frozen into a crust, it is much less affected by the wind. Thus one finds areas where the ground is snow-covered in winter and the vegetation quite plentiful in summer, and other areas which are windswept and therefore quite bare of plants.

In contrast, too much snow can also be a bad thing, for if it gathers into too deep a drift it may not melt at all the following summer. Permanent ice patches result, which on a grander scale are the icecaps and glaciers. Apart from the vast icecap of Greenland, which covers about 95% of the island, permanent ice is not particularly common in the arctic. There are small icecaps on some of the Canadian islands, notably Baffin and Ellesmere, on Iceland and Spitsbergen, and on Franz Josef Land, northern Novaya Zemlya, and Severnaya Zemlya off the coast of Siberia, but elsewhere there are no large expanses of permanent ice though of course small patches occur very widely. Sea ice is formed in winter throughout the northern parts. Sheltered fjords often freeze first and melt later so that close to land there may be ice for longer than out at sea. The presence of ice against the land can have a marked cooling effect which in turn bears on the plant and animal life that can live there.

Arctic vegetation is collectively termed tundra and superficially resembles the heathland of regions further south. However closer inspection shows it to be composed of relatively few species, none attaining any great stature, and all extremely hardy. Although there are no real trees, there is certainly scrub in some areas of the low arctic, and individual bushes may reach ten or twelve feet in height and form quite dense thickets. The predominant species are birch and willow, and they provide a most important habitat for birds in the form of shelter and nest sites. Further north in the high arctic there are no bushes; willow still grows there but instead of upwards, it grows along the ground. Willow 'trees' can extend over several square yards but entirely horizontally, even though a single 'branch' may be several feet in length. The leaves of this and other arctic plants are very small thus reducing transpiration of what little moisture there is.

In the low arctic there are many berry-bearing plants, usually shrubby ones with vernacular names such as bearberry, snowberry, and so on. These are a vital food source for many bird species in the late summer when they are seeking to build up their fat reserves before undertaking the long autumn migration south. Other well represented plant families include the saxifrages, chickweeds, buttercups, and grasses. Many are important foods for the birds, especially the grazing species such as the geese, and also surprisingly for others such as waders which are not usually thought of as plant eating. Mosses are particularly common, and in some areas dominant, especially in the wetter bogs, while lichens abound on the rocky ground, forming brilliant patches of reds and yellows among the many more sombre greys and browns.

Plant species are few in number. For example, in Spitsbergen there are only about 70 different flowering plants, including grasses, while in Iceland, much further south and with a much more varied habitat, the list numbers just over 400. This compares with well over 1000 in an average English county. Part of the paucity is due of course to physical isolation, especially true of islands, but even on the mainland of arctic Canada where there are no geographical barriers to the northward spread of plants, there are still only a few hundred. The conditions are too harsh to allow any more to survive there. In

Purple saxifrage, *Saxifraga oppositifolia,* is one of the earliest tundra plants to flower, often within days of emerging from the snow.

One of the taller arctic plants, the chickweed, *Stellaria longipes,* gains shelter from the wind for its weak stems by growing among stones.

the more barren tundra areas the vegetation is scattered, so that single plants may grow in a field of stones or sand. Here the conditions are so marginal that even these few plants are only just surviving, and although there is plenty of room for more, none will come to fill the gaps.

Propagation is not always by seed. Some form vegetative bulbils which fall off and grow to new plants in their turn. All the flowering varieties are geared to complete their flowering cycle in the very short summer, but if they are still in bud when the first snows of winter cover them they just mark time under their protective blanket, and then as soon as the snow has melted start growing again and come quickly into flower. Thus although they have missed one season they are very well placed to take advantage of the next. Despite the wealth of insect life, few plants rely on them for pollination, most being self-fertilising.

One of the curses from the human point of view of many areas of the arctic is the mosquito. It occurs in what can only be described as plague proportions, making some regions almost totally uninhabitable for periods in the summer. Fortunately they are not malaria-carrying but the bites they give are still unpleasant if only because of the sheer numbers that one is likely to suffer. Eating becomes almost impossible without taking many into the mouth at every bite. They get in the eyes, ears, and hair. They cannot be excluded from tents. They take little or no notice of repellents. A colleague in Greenland was so badly bitten around the face, the only exposed part of his body, that his eyes nearly closed with the swelling, and he ran a temperature from the effect of so much irritant pumped into him by the biting hordes. Even entirely covering all bare skin may not keep them completely at bay as they seem able to bite through thick wool with great ease, and on warm days, when they are at their thickest, wool becomes unbearably hot.

A herd of caribou picking its way across the Alaskan tundra.

So one gets bitten, praying all the while for a cold wind to kill them off, and wondering at the purpose of all this insect life with apparently very little in the way of birds to eat it. Insect-eating birds as such do not occur in the arctic, at least not those species wholly dependent on them. Several do eat insects or feed them to their young but none can afford to rely on them completely. A sudden change in the weather, a day of cold winds, and there will not be a single insect in sight. No bird can afford such a fickle food supply.

Although mosquitoes occur in uncountable numbers, the variety of insect species, as with plants, is very small compared with temperate lands. In most regions there may be only one or two species of butterfly, one bumblebee, a small number of springtails in the tidewrack, a spider or two, and mosquitoes and midges, all adding up to perhaps 100 different kinds compared with the several thousand to be found in a small area of Britain or America. Land mammals too are few. The isolated islands have the least because any land animal living on them has to stay there throughout the year, being unable to migrate south to warmer lands for the winter. Musk ox, reindeer or caribou, arctic hare, lemming, and arctic fox are all that will be found. None except the lemming, and then only in cycles, is very common, though all are comparatively easy to see, for they are considerably tamer than their counterparts further south.

The final feature of the arctic that must be mentioned is the shallow, rich seas which occur around many of the coasts. Some of the most productive areas occur between arctic Canada and West Greenland, around Spitsbergen and Bear Island, and further east around Novaya Zemlya, and also in the Bering Sea. Here plankton grows in summer to form a rich, nutritious soup. Fish undertake long migrations to seek it, and seabird colonies of teeming millions live on the nearest cliffs. And these last are also responsible for bringing a proportion of this nutriment to land where it has a vital effect. They feed themselves and their young on the plankton and small fish, and their droppings are washed down out of the breeding cliffs by rain and melting snow. Wherever flat land occurs at the foot of such cliffs, lush green vegetation is found, fertile beyond compare when set alongside the normal tundra away from the seabird colonies. This rich vegetation in turn provides feeding for grazing animals and birds, forming a unique and important food chain.

2 The adaptation of birds to the arctic

The definition of an arctic bird is almost as arbitrary as the definition of the region. A number of species are very obviously restricted almost entirely to the arctic, scarcely breeding outside it, if at all, and examples of this category are the Brent and Barnacle Goose, Spectacled Eider, Knot, and Ivory Gull. Others are commonly found both inside its boundaries and outside, and the main consideration then becomes whether it is an arctic bird which has spread to other zones, or whether it is a species which typically belongs to another zone, for example the northern boreal, but which has successfully colonised the arctic. Lists of 'arctic' birds have been compiled by various authorities using different criteria, ranging in length from about 80 to 140 species and subspecies. The treatment here falls somewhere between these extremes.

In this book a species or subspecies is categorised as arctic either if it is entirely restricted to the arctic for breeding, or if the greater part of its range lies within the arctic. Where it may have originated from many thousands of years ago is less important than its present distribution. This primary group comprises 95 kinds and to it has been added a secondary group of 37—those which are common in the arctic but which have a much more extensive range outside. These 132 arctic species and subspecies can be divided into 37 waterbirds, 35 waders or shorebirds (which can be classed also as wetland species), 29 seabirds (in the widest sense), and 31 land birds. Thus the proportion of land birds is only just under a quarter, while wetland birds comprise a little more than a half, and seabirds the other quarter. This is quite a different breakdown compared with temperate Europe and North America, where just over 60% of the regular breeding species are land birds while the other three categories of water, wetland, and seabirds each account for between 12% and 14%.

There are a number of interrelated reasons for this great difference. First the habitat diversity of the arctic is much smaller than in the temperate zone, producing many fewer ecological niches for different species. Each niche must provide a nesting site together with adequate food for the adults and for the rearing of their young. The complete absence of vegetation more than a few inches in height, except in a few favoured areas of the low arctic, is very restrictive, as is the small range of foods available. The species that do occur are either heavily reliant on the rich seas surrounding the coasts, or dependent to some extent on the predominantly wet nature of much of the land surface.

The remoteness of the area from suitable wintering grounds is a further restraining influence on species diversity. Some parts, such as Greenland and Spitsbergen, are separated from land to the south by hundreds of miles of sea, necessitating a long and hazardous migration for birds which wish to breed there. And in those regions which are part of a land mass, as in Canada and Siberia, the cold continental winters mean that the

The tree-line is the best biological indicator of the southern edge of the arctic. As they reach the limit of possible growth the trees become more scattered and stunted.

Sea-ice can drift against the arctic coast at any time during the summer. Its presence will have a marked cooling effect on the land.

The coastal plain of west Spitsbergen has patches of relatively fertile tundra bogs and small pools, interspersed with barren stony areas.

Opposite: Lichens form the dominant flora of large areas of arctic tundra, rivalling the flowering plants in variety and colour.

nearest unfrozen wintering grounds may still be hundreds if not thousands of miles to the south. Thus although migration over land may be easier than over the sea, it is still a very long one. This need to make a long journey puts a premium on size and strength and, on average, water and wetland birds are larger and stronger fliers than land birds.

The present composition of the arctic avifauna owes a great deal to the period during and after the last ice age (approximately 10,000 years ago). There is little or no certain knowledge of birds there before the last glaciation but it is known that during it there were several areas free of permanent ice within the total glaciated region of the northern hemisphere. Although the icecap extended south as far as southern England and to the south of the Great Lakes in North America, contrary to common belief it was not continuous all the way from the North Pole. Ice-free areas, or refugia as they have been termed, occurred throughout the various ice ages, and were of enormous importance in allowing all kinds of organisms, plants, and insects, as well as birds, to survive the glaciations more or less in situ, and not be forced to retreat thousands of miles to the south. A number of bird species are thought to have successfully lived through the ice ages in these refugia and then, when the ice retreated, to have spread out from them. Some would have been year-round residents but most would have bred in the refugia and migrated south for the winter, much as they do today.

Other species were pushed south by the encroaching ice and then, when it ceased advancing, were able to live satisfactorily in the tundra lands that formed immediately to the south of the icecaps. They would then have been able to move north again when the ice retreated. These two groups probably form the bulk of the present-day arctic bird

Large flocks of Barnacle Geese winter on the Dutch polders, and on islands off Scotland and Ireland, returning to the arctic to breed.

life being joined by a third, smaller group consisting of those species which, since the last ice age, have been able to move into the region by virtue of their adaptability. However few of these qualify for true arctic status, falling mostly into the secondary category already mentioned.

The areas covered by ice during the maximum glaciation of the last ice age are shown below, together with the main refugia. The most important of these has been named Beringia, after the Bering Strait. During the ice age the sea level was considerably lowered by the taking up of so much water to form ice, and a land bridge formed across the Strait. The refugium covered most of Alaska together with a large part of northeast Siberia, altogether forming an area the size of western Europe. Not only did many species of birds live here during the ice age but the fact that there was a land bridge between Asia and North America had an important bearing on the subsequent distribution of all kinds of organisms. When the ice retreated and the rising sea cut the land bridge, many species found themselves occurring on both sides of the new strait and were therefore in a position to spread both east and west around the Pole. The notable circumpolar distribution of a great many plants, animals, and birds is due to this temporary joining of the two continents.

Some species appear to have found refuge in more than one place. For example the

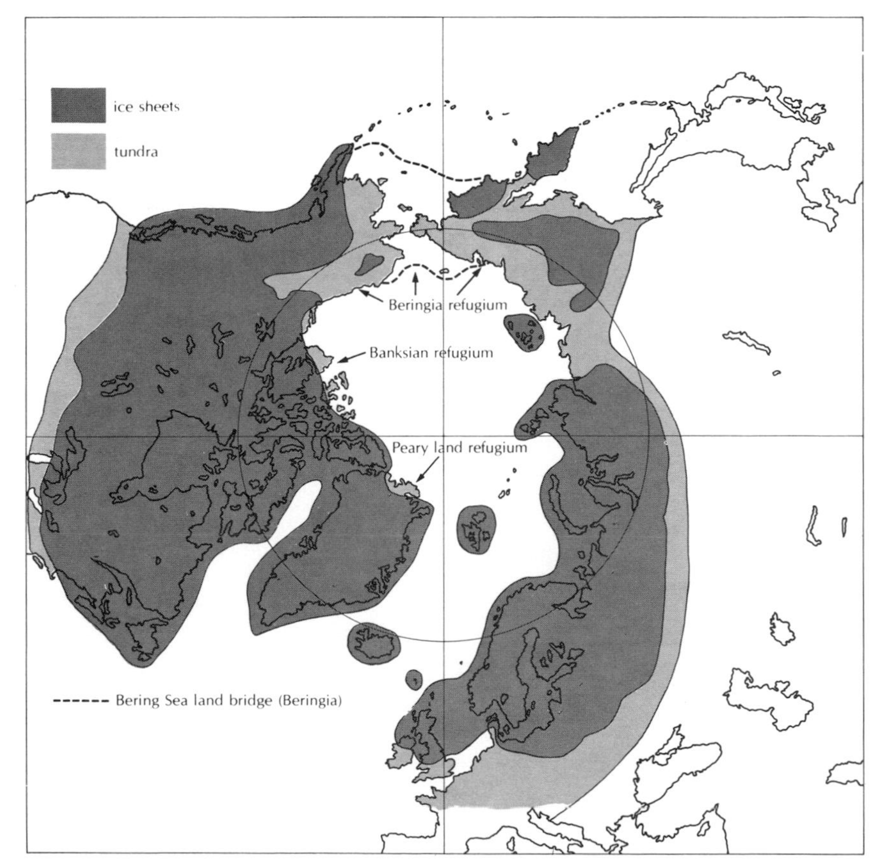

The effects of the last ice age at its maximum, and the approximate extent of the Bering Sea land bridge, which formed when the sea level temporarily dropped.

Opposite: Lesser Snow Geese rising from a marsh in British Columbia. They pause here on their migration from northeast Siberia to California.

Fulmar was almost certainly in the Beringia refugium but its present distribution suggests that it also survived somewhere within the North Atlantic area, probably to the south of the main icecaps there. Similarly the Snow Goose occurred in Beringia but also lived in the Banksian refugium, located approximately around today's Banks Island in arctic Canada. However being much smaller, many fewer species probably lived there than in Beringia. This is even more true of the Peary Land refugium which was the most remote and probably also had the least hospitable climate. One of the birds that managed to live there was Hornemann's Redpoll, which is still the most northerly-living small bird. It is supposed that it migrated south for the winter, but some may have been able to stay there all year.

In addition to these completely enclosed refugia there were the tundra areas, sometimes also called refugia, to the south of the

icecaps. These were 'islands' in the sense that to the north was ice and to the south forests or steppes. However life here would certainly have been easier for birds because of the more ready access to warmer wintering areas with no requirement for a lengthy flight. It has been suggested that the icecap prevented birds from reaching the refugia, but this ignores the fact that a number of species today make regular migrations of several hundred miles over the Greenland icecap, having to climb to over 10,000 feet in order to do so.

Virtually all the 95 truly arctic species dealt with must have been arctic dwellers during the last ice age and ever since. Their distribution has altered as the extent of the icecaps and therefore the fringing tundra habitat has changed, but their arctic origins are not in doubt. The 37 secondary species may not have been arctic birds during the ice age but have certainly become so since. They have managed to adapt to the somewhat restricting circumstances of breeding there, either because of some inherent ability to fit into a variety of situations, or because some outside pressure on their existing habitat has pushed them northwards to survive if they can, or to perish if not.

The arctic region is far from ideal as a place for birds to live, even just for the summer as with the great majority. Very few indeed manage to live there all the year round and those that do have had to adapt to the rigours of the climate. The remaining migratory species also labour under some disadvantages, not only the distances they have to travel to reach their breeding grounds, and even in summer there still remain several drawbacks to a comfortable life. The major hazards that birds encounter are three-fold. First is the cold: this principally affects the resident species but can also be a problem for the migrants at the beginning and end of the summer. Secondly the summers are very short, placing a number of restrictions on breeding. And thirdly, related to the habitat and climate, the availability and abundance of food is by no means as secure as it is in more temperate lands.

The most widespread land predator is the arctic fox. They feed abundantly in summer but suffer very lean times in the harsh winter.

Counteracting these disadvantages there is one major advantage for the birds—the very remoteness and winter inhospitability of the land. Although primarily an adverse factor, it means that much of the area is uninhabited or only very thinly peopled, so that disturbance and exploitation by man are greatly reduced. What is more, land predators such as foxes, wolves, and other carnivores can only themselves survive if they are able to find sufficient food to see them through the winter, so therefore either they too must migrate like much of their food or, where they live on islands, the winters will provide a severe check on their numbers. Bird predators are also fewer than further south as each one has to be capable of a long migration.

The first physical condition that most people think of when the arctic is mentioned is the cold. The ways that the birds have adapted to overcome this problem can be divided into those which have produced changes in body size or in the relative proportions of parts of their body, and those to do with insulation, either physiological or involving specialised behaviour. The conditions that birds have to cope with naturally vary from place to place but clearly they have to be able to stand up to the worst that nature can throw at them. Although much work has been done on the problem, there are still unanswered questions, and even some of the answers are open to doubt. For example it has been shown experimentally under laboratory conditions that birds are capable of living for several hours in very cold temperatures, down to as low as —100°C (—148°F), or considerably below anything they are going to meet in the wild. But attempts to move from this result to natural conditions must be treated with the utmost caution. In captivity a bird has its food provided and need expend only the very minimum of energy in eating it, conserving all the rest for maintaining its body temperature. In the wild a bird has to expend a lot of energy in finding its food and this will make large inroads into the energy left for keeping warm. It must also be borne in mind that in the wild the cold temperatures of winter are also associated with a minimum of daylight during which feeding can take place. So although it may be physiologically possible for a bird to survive very low temperatures it does not mean that the cold is no problem.

Bergman's Rule, named after the Swedish scientist who promulgated it, states that birds breeding in the north tend to be larger than their counterparts further south. Although there are exceptions, this is true for a wide variety of species. An increase in the size of an object reduces the area of the body surface relative to its volume so that the larger the bird the smaller, relatively, is the surface available for heat loss. Examples are common in the arctic, among them the Wheatear and Ringed Plover whose subspecies breeding in northeast Greenland are about 20% larger and heavier than the Icelandic subspecies, which in turn are larger than those in the British Isles. Differences are even more marked in seabirds, perhaps because the conducting capacity of water is very much greater than the air, although this must be offset by the fact that the lowest sea temperature that can be experienced is only —3·5°C (25·7°F), below which of course it freezes. Further examples of Bergman's Rule will be found in the species accounts in the following chapters. There are however disadvantages for birds in being too large, notably that it becomes difficult to find sufficient food both for themselves and for the successful rearing of their young in the short summer, so that after a certain size Bergman's Rule ceases to operate, and even goes into reverse. Thus of the four northern swans, the North American Whistling Swan and its Eurasian counterpart the Bewick's Swan are smaller and breed further north than the other two, the con-specific Trumpeter and Whooper. Similarly among the multi-raced Canada Geese the smallest are found further north.

Another common method of cutting down on heat loss is the reduction in the size of bodily extremities, especially bills in the case of birds, and ears, nose, and tails in mammals. This reduction in extremity size can clearly be seen in the Eskimo. In birds there are examples to be found in a number of species, including Brunnich's and Black Guillemot, Fulmar, and Arctic Tern.

Insulation has been improved in a number of ways. Several arctic species are white or have extensive pale areas, and although there is a distinct camouflage advantage, it is also true that white feathers are better insulators than coloured ones. This is because the pigment in a feather lies within its structure so a

An adult and an immature Whooper Swan. The young bird will not breed until it is at least three years old.

white one (which totally lacks pigment) remains hollow, and as is well known any air space is an effective insulator. Furthermore a white object radiates less heat than a coloured one. The camouflage aspect of white feathers may not be as important as the insulating properties because a number of species are white or nearly white not just in winter when snow is on the ground but throughout the summer as well when it might be thought to be a positive disadvantage to be so conspicuous. There is also a tendency for some birds to be paler in northeast Siberia than in the northwest, not because there is any more snow there but because it is distinctly colder. Additionally, certain species have white areas which are not particularly visible nor used in display, so their only likely function is insulation. Examples of this include the Ptarmigan, Black Guillemot, and Snow Bunting. Although the Ptarmigan moults from its white winter plumage into a brown summer one, in the very short high arctic summer it rarely actually completes the spring moult before it is time to moult back again into white, always before there is any snow lying on the ground. And the first area to turn white is on the underside which would be the most important in terms of heat loss.

The feathers of arctic birds are thicker and more numerous than those of their southern relatives. In particular the downy bases of the body feathers are denser and therefore warmer. They are also used as insulation on the legs of the Snowy Owl and Ptarmigan, two of the four species of land bird regularly

The white winter plumage of the Willow Ptarmigan is excellent insulation against the cold as well as perfect camouflage.

Opposite: The legs and feet of the Snowy Owl are feathered to protect them from the cold.

wintering in the arctic. The other two species, Raven and Snow Bunting, do not have feathered legs but both have been observed keeping their legs warm by walking in a half crouching position so that the upper parts were obscured by the belly feathers which in turn were lowered slightly. When sitting on snow or ice, birds always tuck their legs and feet well into their belly feathers, and some long-legged birds may even draw their legs up when in flight instead of letting them trail behind as usual. This behaviour was observed among waders in Britain during the very cold weather conditions of the 1962/1963 winter.

There are other behavioural traits adopted in very cold conditions which are not normally seen. On surfacing after a feeding dive, diving ducks such as Eiders shake their plumage in order to rid it of drops of water that might otherwise quickly freeze in the cold air. As already mentioned unfrozen sea water cannot go lower than −3·5°C (25·7°F) whereas the air temperature may be very considerably colder than that. On coming out onto land, waterbirds will preen very quickly and thoroughly to rid all their plumage of water droplets, because the freezing of even a small amount of ice onto their feathers would soon stick the plumage together and cause it rapidly to lose its insulating properties.

The short summer entails a certain amount of compression, certainly by the larger species if they are successfully to complete the breeding cycle within the available period. Birds such as divers, swans, and geese arrive

on the breeding grounds already paired, and often before the snow has cleared from the ground. They are physiologically ready to move straight to the first snow-free nest site, and they start to lay eggs with the minimum of nest preparation. The geese and swans feed as much as they can before and during the spring migration, laying down not just enough fat reserves for the journey, but sufficient to survive in case there is no snow-free ground or vegetation to eat for a period after their arrival and, for the female, sufficient for the production of the eggs. Their incubation and fledging periods are shorter than in similar-sized species nesting further south, with the young having a much faster growth rate enabling them to reach the flying stage as early as possible. Then there is a short period of two or three weeks at most in which the young have to become strong on the wing before departing, by which time it is likely that the temperatures are again falling below freezing and snow is beginning to lie.

Ross's Goose is one of the smallest geese. It protects its nest from arctic foxes by siting it on an island in one of the lakes.

A Barnacle Goose on its nest in Spitsbergen. This species breeds wholly within the arctic and migrates over 1,000 miles to reach its winter quarters in north-west Europe.

Studies in northern Canada have shown that in an average year there is a frost-free summer period in the Perry River region of 93 days. The Ross's Geese that breed there require no less than 80 days of this period to complete their cycle, from the selection of the nest site to the point when the young are strong enough on the wing to migrate. Clearly this leaves very little leeway, particularly for a late spring, and it seems certain that rather than run the risk of still having unfledged young when winter returns, the birds will not even make a start rather than attempt to breed late. If on arrival they find completely unsuitable conditions they will only stay in breeding condition for a comparatively short period, probably no more than two weeks, before physiological changes take place which effectively end the possibility of reproducing that season.

The phenomenon of non-breeding has received much attention since it was first noticed thirty or forty years ago, but it has come to be recognised as a normal part of life in the arctic. Also, some earlier workers failed to realise that for certain species such as the geese and swans there is a not inconsiderable proportion of immature, non-breeding, one and two year old birds present every year. But even so it is clear that occasionally a very late spring will completely inhibit any attempt at breeding in a wide variety of species and over a very large area. Of course such seasons do not occur too often otherwise populations would die out, but they form a very serious hazard and in some places may occur as often as two years in five. Originally in order to detect a non-breeding year it required observations actually made in the arctic, and it was this that led to the supposition that they were rather rarer than has

turned out to be the case. But in the last twenty years or so it has been regular practice in both North America and Europe to monitor the breeding success of arctic goose and swan populations by making sample counts of the percentage of young birds in the flocks as they pause at autumn migration resting places or reach their winter quarters. These have revealed that for high arctic nesting species such as the Brent Goose, complete failures to breed are nearly as frequent as successful years.

It is not certain however that a breeding failure as detected in the following autumn is always due to a late spring. The need to compress the breeding season and thus to make as early a start as possible means that birds of the same species will begin to lay their eggs more or less simultaneously over a wide area. Indeed most of the pairs in a population of geese may lay their eggs within a matter of one or two weeks. In more temperate lands the laying season would be extended over two or three times this period. Although simultaneous laying is necessary if the birds are to complete their breeding cycle successfully it also means that any spell of bad weather at a critical stage will affect almost all the birds equally. A period of cold winds and rain or snow at the time the young are hatching can cause very heavy losses and convert a good breeding season, when an early spring has enabled the majority of pairs to find nest sites, into a complete failure. For in the arctic there is no time to lay a replacement clutch.

Although the larger birds like geese and swans are most affected by late springs and by bad weather later in the season, other species can also have great variations in annual breeding success. A study in Canada showed that a five-day spell of near freezing temperatures, strong winds, and heavy rain killed nearly 90% of wader chicks and nestlings of land birds. Those that survived were the ones still being incubated as eggs which are much less vulnerable. Most small birds place their nest either in rock crevices or as deep into thick vegetation as they can. In addition to this natural insulation and protection they line their nests with feathers and even cover the eggs partially with them. As a consequence, the temperature in the nest is hardly affected by what happens outside and this period at least of the breeding cycle is relatively untouched by bad weather.

The continuous daylight of the arctic is a positive boon to the parents as it enables them to find food throughout the 24 hours. Studies have shown that although they have a roughly 24-hour rhythm their spells of sleep or inactivity are very much shorter than in related species further south where it gets dark at night. They spend a far greater proportion of each day gathering food for their chicks, or in the case of nidifugous young, feeding themselves. (Nidifugous young are those that leave the nest on hatching, as opposed to nidicolous—those that stay in the nest and are fed by their parents.) This means that the chicks grow much faster and thus cut down the time spent in the nest or in growing to adult size and full flight capability. It is a good example of how the breeding season has been successfully shortened by the birds taking advantage of one arctic characteristic, the continuous daylight, in order to overcome another, the short summer. The ability to feed almost throughout the 24 hours is taken advantage of in a different way by some of the smaller birds, among them the Redpoll, Snow Bunting, and Wheatear. In the arctic races of these species the average clutch size is larger than in their southern counterparts. The female can use the continuous daylight first to find enough food to make the energy to lay more eggs, and then, with the help of the male, enough food to rear the larger family. This increased productivity is no doubt important to offset the greater losses to be expected in bad years or on the long migrations. In addition arctic land birds generally have a high hatching success.

The Wheatear of high-arctic Greenland is larger and brighter than the bird of temperate latitudes. The Iceland race is intermediate in size.

At the end of each breeding season it is normal for all birds to change their plumage, to moult out the old feathers and replace them with a new set. Some have a partial spring moult as well. The post-breeding moult takes up several weeks or even months, depending on the species, and requires a considerable amount of extra energy in order to grow the new feathers. It is usual for summer migrants breeding in temperate zones to carry out their moult between the end of the breeding season and the start of the autumn migration, but arctic birds just do not have long enough to do this, and have had to find ways round the difficulty. In some it starts while the breeding season is still under way, so that as well as rearing a brood the adult has also to grow new feathers, a considerable extra strain. The moult itself is accelerated so that instead of dropping at most a couple of wing feathers at a time and waiting until the new ones are growing well before shedding any more, they lose four or five at a time and have to accept the consequent reduction in flying ability. Wildfowl of course are already well adapted by reason of their method of shedding all their main flight feathers simultaneously so that the period is reduced to a minimum. The fact that they are flightless for three to four weeks is no problem—they can be safe on water even though they cannot fly. Another way of coping is to delay the moult until after the autumn migration or, as in the case of some waders, to start it and then to arrest it while some or all of the migration is completed, before restarting it much further south where there is no danger of winter overtaking the birds.

Yet another answer, which although not confined to arctic species is of great importance to some of them, is moult migration. This is carried out by non-breeding birds, at least in the arctic, and involves them in a special migration away from the breeding grounds to some separate area where they stay for the duration of the moult before embarking on the normal autumn migration. It is more or less confined to wildfowl and is most prominent among some of the geese, though it is also found in several species of duck. Typically, in the geese, it involves a northward journey to an area where that species does not breed. It has been very plausibly suggested that by removing the non-breeding component of the population (which may be quite large in a goose species because it includes one and two year olds) from the breeding grounds, the maximum amount of food is made available for the parents and young. The northward movement, sometimes of several hundred miles, takes the birds to places that may be quite suitable for them to live in for short periods but that do not have a sufficiently long summer for the completion of the full breeding season. The moult migration thus exploits as far as possible those areas that would otherwise be unused by that species.

The food available in the arctic may be rather more restricted both in variety and quantity than further south. In particular, the birds have to be adaptable in their diet; there is little room for specialists. Although insects such as mosquitoes abound at certain times their presence is governed by the right weather conditions. Consequently there are no birds completely dependent on them, and among those that prefer an insect diet, all without exception are capable of living on other things, especially plants, should the insect supply fail. In winter all the wader species live almost exclusively on invertebrates of some kind, mostly found by probing in mud and water. In summer this food is also available to them in some quantity, including as it does mosquito adults and larvae. However on the birds' arrival in spring the insect life is in short supply and at this time, and in spells of bad weather during the summer, they must live on the leaves and seeds of plants. Similarly the small land birds feed on insects when they can, especially taking them for feeding their young, but all are also able to feed on seeds and other vegetable matter. Two of the species that over-winter in the arctic, the Ptarmigan and the Redpoll, have highly specialised adaptations for feeding: they have developed crops, or crop-like organs, in which they can store food. This means that they can take in more than they can immediately deal with and thus achieve a higher total intake for subsequent digestion and conversion into energy.

The only assured food supply is in the sea and it is noticeable that the seabirds are by far the most numerous. Non-breeding among them is apparently unknown, another pointer to the more stable conditions they experience.

Unlike their relatives further south, the Mallard of Greenland do not breed until they are two years old.

Nonetheless even they have had to adapt to some extent, most obviously by a shortening of their breeding season, accompanied by more rapid growth of their young, which are reared on their abundant sea food supply.

Among many arctic birds there is a tendency to delay the attainment of maturity. Thus the Mallard in Europe and North America commonly breeds in its first summer after hatching, when it is approximately eleven months old. Indeed a few precocious individuals have been found breeding in their first autumn when only 6 to 7 months old. The Greenland Mallard, however, a distinct subspecies living in the southwest of that

country, does not breed until its second summer or just before it is two years old. This delay, also found in other species, means that when the birds do breed for the first time they are that much older and more experienced at finding food and looking after themselves. A totally inexperienced bird breeding for the first time in poor conditions would have a much higher failure rate and might even die itself in the attempt, which would be bad for the species as a whole.

Migration has been mentioned a number of times. The nature of the arctic winter makes this phenomenon virtually compulsory for all but the hardiest few species. The freezing of both fresh and salt water and the depth of snow prevent the birds from finding their food, and so they are forced to fly to areas where they can find it. For most this means at least far enough south to escape freezing conditions, while many kinds, particularly the waders, fly on across the equator to spend their winter in the summer of the southern hemisphere amid an abundance of insect life. The few that remain in the arctic, Ptarmigan, Arctic Redpoll and, in a few areas, Black Guillemot, Snowy Owl, and Raven, can only survive because of two important factors. First their food is still available: mainly seeds on vegetation remaining above the snow for the first two species; fish and other life in places where open leads occur in the ice for the Black Guillemot; lemmings and voles for the Snowy Owl; and a wide variety of items for the scavenging Raven. And secondly their numbers are always small so that what little food there is can still go round.

The northward migration in spring has to be timed so that the birds do not arrive on the breeding grounds too soon. The larger geese and swans must start their egg laying while the ground is still largely snow-covered, living on their accumulated fat reserves meantime, but the ducks, waders, and smaller birds need to be able to feed as soon as they arrive. Each species has its own pattern both of timing and method of migration. The waterfowl move in large flocks, the pairs already formed, and make a number of long flights calling in at regular stopping places on the way. Waders, too, by virtue of their restricted habitat requirement of wetlands, have their traditional halts and generally move in flocks. However in many species pairing does not take place until the birds reach their breeding grounds, and so some flocks will be virtually all males, while others will be made up of females. Both waterfowl and waders usually fly at night, several thousands of feet above the ground or sea, where they encounter less turbulence than lower down and are also free from the possibility of flying into high ground. The migration of smaller birds is less well known, mainly because they do not concentrate at specific localities but are able to travel on a broad front, moving forward steadily but unspectacularly. Their progress on land is akin to that of seabirds on the sea, where the habitat is continuous and the birds are able to pause in their northward movement more or less whenever they wish or whenever weather conditions become unsuitable.

The southward migration in autumn, like that in the spring, is triggered initially by hormonal changes under the stimulus of changing daylight length as the seasons advance or retard. But for many species the actual moment of departure will be forced upon them by the more proximate factor of weather. Autumn hardly exists in the arctic. The summer has barely ended before winter begins, and it can begin very abruptly. In some localities freezing conditions and heavy snowfalls often occur in early September, causing the birds to depart at once, but if the weather remains open then the birds will stay on for a further period. There is great variety in the behaviour pattern. Some migrate as families, for example the divers, geese, swans, and some seabirds. With the exception of the divers, which probably always remain as separate family parties, the birds from a breeding colony or a wider area will gather in flocks in the weeks prior to departure. If the terrain allows, they may move gradually southwards but often they stay on or near the breeding grounds until the moment of leaving. This, if weather-influenced as described above, can be a spectacular affair with thousands of birds on the move together. Nearly all other species—ducks, raptors, waders, and smaller birds—lack strong family bonds and the young are deserted by their parents at or soon after fledging, or in some instances even before they can fly. The adults and young migrate separately, mostly in flocks in the case of waders, or singly or in small groups for the others. The young birds have to rely entirely on instinct to take them in the right direction and the correct distance to their winter quarters. This is very different from the young geese, for example, which merely follow their parents to the right destination.

Migration is typically considered to be the complete removal of a population of birds from a breeding area to a totally separate wintering area. Most arctic species are complete migrants, with the whole population leaving the nesting grounds for a wintering area further south. But several, like the Snow Bunting and the Eider, are partial migrants; some movement takes place but there is an overlap between the south of the breeding range and the north of the wintering range. In such cases there are two possible ways in which the population moves. It may either be a complete shift southwards by all the birds or, alternatively, those in the south of the breeding range may remain where they are while those from further north migrate over them to the wintering area further south. This is called 'leap-frog' migration and has been detected in a number of species, including the Canada Goose, Common Eider, and Redshank.

The distances travelled and some of the routes taken by migrating birds are astonishing. Several species of waders breeding in arctic North America spend the winter in the south of South America, having flown the complete length of the continent. Those from arctic Eurasia also fly as far as they can, in many cases reaching Australia and Cape Province in South Africa. A number of birds have colonised the northeast of Canada from Europe, as is made apparent by their migration route, which is southeast to Europe, before heading south to Africa. This prodigious journey is accomplished by, among others, Knot, Turnstone, and Wheatear. They, together with Brent and Greenland White-fronted Geese, have either to make a detour round or a hazardous flight over the frozen mass of Greenland lying directly in their path. Several hundreds of miles across and up to 10,000 feet high in the centre, this forms a considerable obstruction. It seems certain that the geese and very probably the waders can and do fly across it regularly, but other birds, including the Wheatear, move south down the west coast of Greenland to Cape Farewell and then make the long sea crossing to the British Isles in preference. A possible alternative would be to follow the north coast of Greenland before turning south down the east side, and thence to Iceland, but observations are, not unnaturally, lacking.

3 Waterbirds

The importance of water and wetland habitat in the arctic cannot be overstressed. Water, both fresh and salt, surrounds small islands and makes them safe nesting places from which land predators are excluded; water encourages the growth of plants, and is the home of insects, plankton and fishes; and swimming birds can seek sanctuary on water, which is of vital importance to wildfowl when they are flightless during their annual moult. The waterbirds dealt with here are the divers and the wildfowl—swans, geese, and ducks. All are swimming birds with webbed feet but the geese feed mainly on land, resorting to water to roost at night and when disturbed during the flightless period.

Divers [Loons]*

The four species of diver are essentially aquatic birds, only coming to land to nest. Their powers of locomotion on land are so restricted that the nest is always placed either in shallow water or within a foot or two of the water's edge. Expert divers, they feed almost exclusively on fish caught by underwater pursuit. All nest on freshwater bodies but in winter stay mainly on the sea where they can be found, usually solitary, in shallow bays or, occasionally, wide river estuaries. The nest of all four is a low heap of waterweed with a shallow depression in the top. The normal clutch is two, rarely three, eggs, and the incubation period is about four weeks.

An adult Black-throated Diver [Arctic Loon] feeding its young with a small fish. As usual, the nest is very close to the water.

* American species names are given in brackets where they differ from the British name.

Red-throated Diver [Red-throated Loon]
Gavia stellata
The most widespread and common of all the divers, the Red-throat is also the tamest. It breeds on the smallest tundra pools, often no more than twenty to thirty yards across and a few feet deep. Clearly these cannot contain sufficient fish, so the birds have to fly to larger lakes or to the sea for their food. The use of such places for breeding removes the possibility of competition with other divers, which all nest on larger waters. In order to live on such small pools it has the ability to take off from a very short run and can even take off from land, which none of the others is thought able to do. The nest is often sited in shallow water rather than at its edge, and if disturbed, the incubating bird will slip into the water and then quickly take off and fly round until the danger is passed. When it has young the parent usually stays with the chicks, swimming up and down anxiously and sometimes allowing an approach to within a few yards. Like many arctic birds this apparent tameness is formed from the desire to protect its young coupled with a certain indifference to man.

The need to obtain the food from somewhere other than the breeding pond entails regular flights to and from a feeding area. In coastal parts this is always the sea, the birds sometimes flying several miles to reach it. Not only do they get their own food this way but they bring back in their bills sufficient for the young as well. One parent usually stays with the chicks, at least while they are small, while the other is away fishing. Later both make frequent journeys to and fro carrying food. On arrival back at the breeding pond the adult usually announces itself with a repeated 'tuk-tuk-tuk'. The birds have a wide repertoire of wails and cries, but the finest are heard from a courting pair and carry for long distances over the tundra. These have the effect not only of firmly establishing the pair bond between the adults but of declaring that this particular pool is inhabited for, like the other divers, they hold territories, usually the complete pond or lake, and do not allow others of their species to land on it.

The Red-throated Diver breeds further north than any of the other species, occurring in the far north of Greenland and in the northern islands of the Canadian arctic. In these regions it cannot nest until the shallow freshwater pools have become ice-free and consequently it lays its eggs up to a fortnight later than the birds that breed further south, often not until the middle of June. However the fledging period of its young, at six weeks, is shorter by an equivalent amount. In winter it is more often seen close to land than the other divers and may also occur in quite large flocks—as many as several hundred on migration. It also regularly winters further south, reaching California, Florida, Portugal, and Japan.

Black-throated Diver [**Arctic Loon**]
Gavia arctica
The Black-throated Diver is about the same size as the preceding species, and although their breeding ranges overlap, this bird does

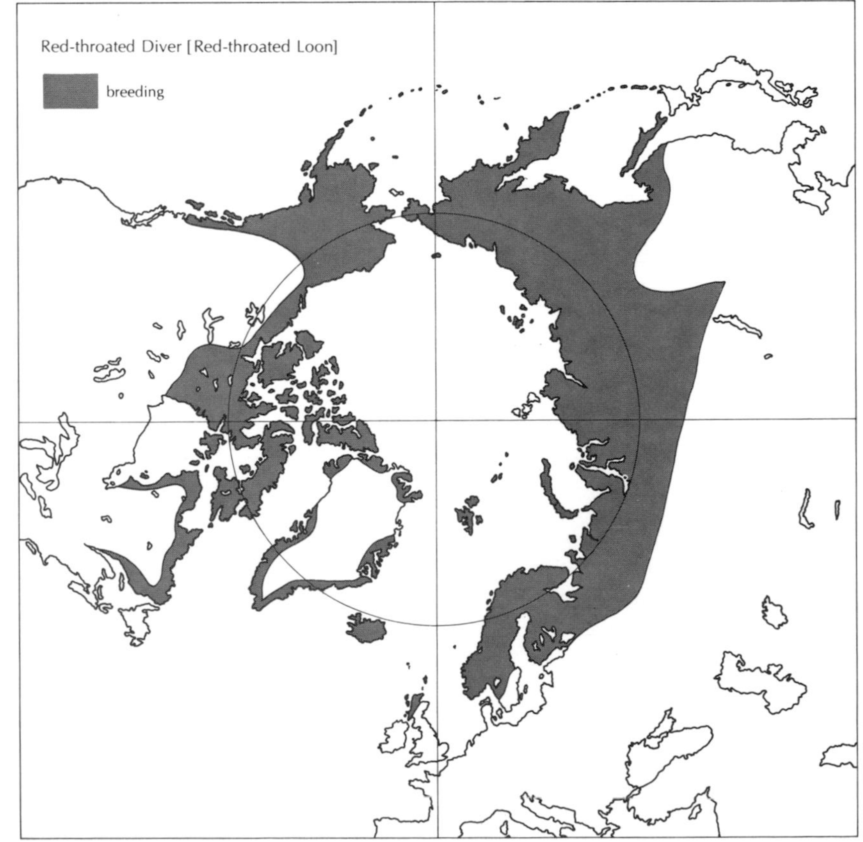

The breeding range of the Red-throated Diver [Red-throated Loon].

not extend nearly so far north. In winter it is relatively common off the coasts of Europe, and loose flocks occur in shallow waters such as the Black Sea. It nests on freshwater lakes, both in the arctic tundra and among the forested areas further south, which have to be large enough to supply all the food requirements of both adults and young. The latter fledge in about two months.

Like all divers, it can submerge without the trace of a splash. By compressing its feathers, thus squeezing the air out from between them, and deflating its lungs, it can sink slowly from view, swimming if necessary with little more than the head showing. Then with scarcely a ripple the head goes down and the bird has dived. Up to a minute is the normal duration for a dive, but as long as three minutes has been recorded. If swimming and diving reveal the bird in its element, flying is much more laborious, particularly the take-off. In calm conditions it may take three or four hundred yards of splashing over the surface with frantically beating wings and paddling feet before it becomes airborne. However, once in the air, flight is fast and direct with the rather narrow pointed wings beating rapidly.

Great Northern Diver [Common Loon]

Gavia immer

One of the most blood-chilling, yet thrilling and evocative sounds of the arctic is the calling of a courting pair of divers, and of the four species the Great Northern has the weirdest yet finest voice. A mixture of yodelling, laughing, and wolf-like howling, the cries echo and re-echo across the breeding lake. This is always large and freshwater, and the pair return here as soon as the winter's ice begins to melt. The nest is usually the typical pile of waterweed but sometimes there is hardly any nest material, just a shallow depression on the bare ground, and it is always placed very close to the water's edge, either on an island or at the tip of a promontory. From it the bird can just slip quietly into the water at the approach of danger, submerging immediately to surface perhaps fifty or a hundred yards away, and swimming low with little more than the head showing.

The chicks are covered in down and are able to leave the nest within a few hours of hatching. They are looked after by both parents, who catch fish for them to eat, small to start with but becoming larger as the birds grow. The fledging period is about two and a half months which, together with the incubation period, takes up just about the whole of the arctic summer. From eggs laid in early June, the chicks will only just have fledged by the middle of September, the beginning of winter. On leaving the breeding lake the family move to the coast and from there migrate southwards to their wintering area. It is probable that the young stay with their parents for part of the winter, but they certainly become independent of them before the following spring. Then the adults return north to their breeding place, but the young birds, which will not be old enough to breed for another one or two years, remain on the sea in or a little to the north of their wintering area.

A family party of Great Northern Divers [Common Loons]. The adults are in their drab winter plumage, and somewhat resemble the juvenile (nearest bird).

The breeding ranges of the Great Northern Diver [Common Loon] and the White-billed Diver [Yellow-billed Loon].

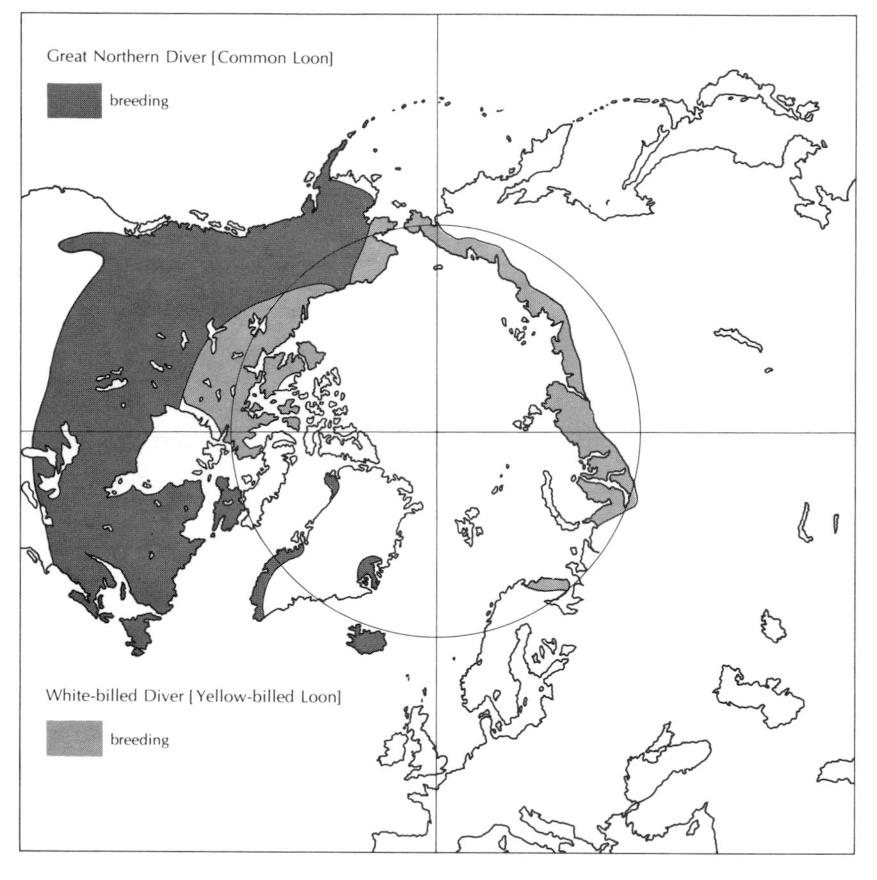

Opposite: The distribution of the White-billed Diver [Yellow-billed Loon] is circumpolar. It is seen here nesting in Alaska.

Great Northern Divers are nowhere common, despite their American name. Persecution in the past, for their feathers or to protect fisheries, has reduced their numbers, and modern disturbance and pollution of the water has prevented much recovery—many wintering birds are washed up oiled on the beaches of Europe and North America.

White-billed Diver [Yellow-billed Loon]
Gavia adamsii
This bird is confined far more to the arctic than the previous species and has an almost complete circumpolar distribution. It is not common anywhere and is relatively little known. It appears to have rather similar habits to the Great Northern but as their ranges hardly overlap there is no competition between them. Possibly they are merely subspecies, although hybrids or intergrades are unknown and so separate treatment seems justified. Their breeding requirements are also similar except that the White-bill seems prepared to nest on smaller lakes and to fly to other waters, including the sea, to find food. However it does not apparently breed on the really small ponds used by the Red-throated Diver. The winter distribution is not well known. Single birds are occasionally seen on the coasts of Europe and North America, but because separation in the field from the Great Northern Diver is difficult, their true status is hard to ascertain. The only regular wintering areas appear to be the coasts of northern Norway, and from Alaska south as far as British Columbia.

Swans

The swans are the largest birds living in the arctic and consequently have had to adapt in several ways in order to complete their breeding cycle in the short summer. They must feed as much as possible in the spring to lay down sufficient fat reserves both for the migration north and, if necessary, to live on after arrival on the breeding grounds when their plant food may be in short supply. In addition the female must have the extra reserves available for the production of eggs, and it is perhaps not surprising that the average clutch size is less than for the species breeding further south. The eggs themselves, though, are large so that the cygnet is particularly well developed on hatching, but thereafter the young must grow at a very fast rate, achieved thanks to the continuous daylight allowing virtually round the clock feeding.

The adults defend a territory around the nest that contains sufficient food for the rearing of the young. The cygnets feed initially on a high proportion of insect food but can also take plant food right from the start as an alternative, and they gradually change to an exclusively vegetarian diet. Their pure white plumage acts as a most conspicuous demonstration of the birds' presence and therefore of the existence of a breeding territory. To this excellent visual signal, also of great importance in keeping flocks together during the winter, the arctic swans add a loud and far-carrying voice.

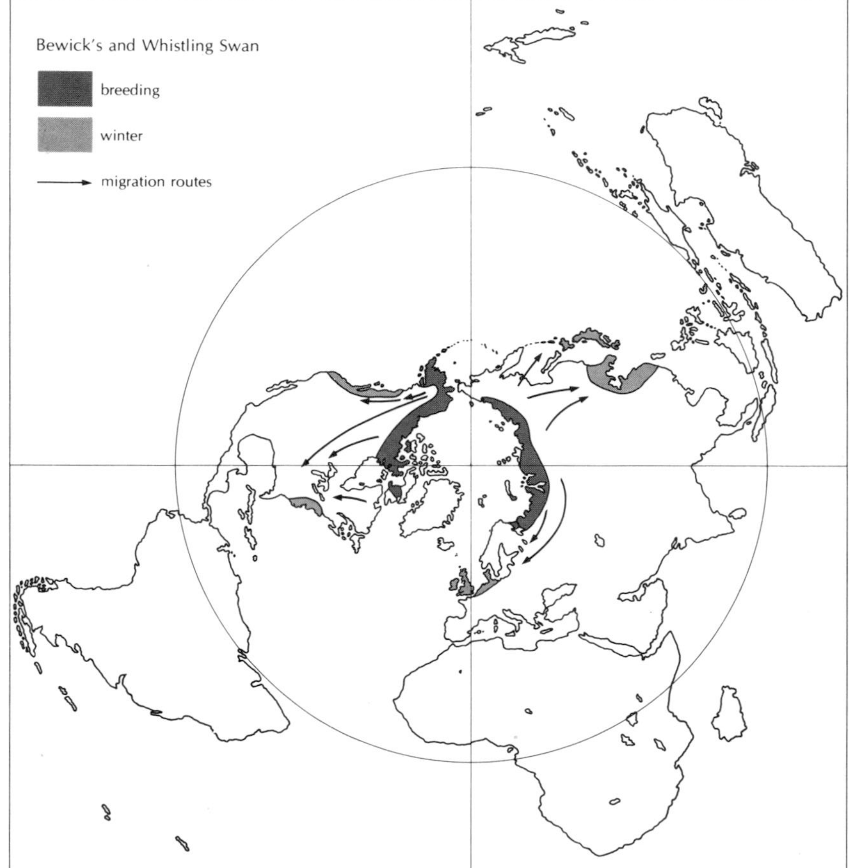

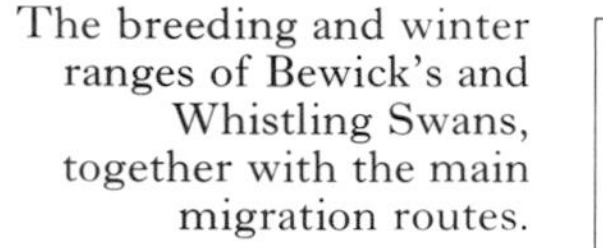
The breeding and winter ranges of Bewick's and Whistling Swans, together with the main migration routes.

Bewick's Swan and **Whistling Swan**
Cygnus columbianus

The only significant difference between these two swans is the far more extensive yellow on the bill of the Bewick's, which is either very small or absent in the Whistling Swan, and they are usually regarded as subspecific, the first breeding in Eurasia, the second replacing it in North America. Both birds build a large nest of vegetation with a shallow depression in the top beside shallow pools in the tundra. The usual clutch size is four or five, rarely six. The female alone incubates, though the male will sit on the nest while the female is away feeding, and the incubation period is about thirty-two days in the Whistling Swan but up to three days shorter in the Bewick's. Within twenty-four hours of hatching the young are ready to leave the nest, and are accompanied and protected by their parents. The adults do not actively feed the cygnets but pull down vegetation to within their reach, and also stir up the water to bring insects and other food particles to the surface.

The young grow apace and although the exact fledging period has not been recorded it is probably around six to seven weeks, little if any longer than some of the arctic geese. The family depart on autumn migration together, joining with other families before leaving, and meeting more at some of the traditional stop-over places on the way south. Both types have an overland migration and can therefore stop and feed at various localities on their journey. The wintering areas are largely coastal: the Whistling Swan occurs on tidal inlets and estuaries in British Columbia and California, on the Pacific coast, and from Maryland to North Carolina on the Atlantic seaboard, and the Bewick's Swan winters on floodlands and estuaries in Northwest Europe from Denmark to northern France and the British Isles, while in eastern Asia they are found in Japan and China.

Whistling Swans have been intensively studied in recent years, especially in order to plot their migration routes. Although they winter on both sides of the continent of North America, the dividing line between the migrants to east and west is nowhere near the centre of the arctic breeding grounds. Instead it is in Alaska, very far to the west; thus some birds are migrating from northern Alaska right across the continent to winter on the Atlantic coast. Accurate knowledge of the routes they take and the timing of their movements is of great importance in avoiding collisions with aircraft. The main danger comes when the aircraft are at relatively low levels, for example when landing or taking off, and better knowledge of the swans' movements would enable aircraft to be re-routed at peak migration times. Censuses put the population at rather more than 120,000: of these up to 60% winter on the east coast, the remainder on the west.

In contrast Bewick's Swan is a much less common bird. The north-west European population is probably no more than 7,000, the majority of these wintering in the Netherlands and the British Isles. The two largest flocks in England have been encouraged to remain on refuges during the winter by feeding them grain. Normally they prefer to graze from low-lying wet fields, but drainage is fast reducing such habitat, and this dietary supplement is seen as a way of aiding a potentially threatened species.

One of the wintering sites in England is at the headquarters of the Wildfowl Trust, at Slimbridge, Gloucestershire. Here up to 400 birds come each day to a small pool within the enclosed area where the captive waterfowl are kept, and close observation of them is possible from behind large windows. Sir Peter Scott, Honorary Director of the Wildfowl Trust, discovered several years ago that he could identify individuals at close range from the varying patterns of black and yellow on their bill, and so began a fascinating long-term study of individually known birds from among a relatively small population. A daily register of the swans is kept throughout the winter, from the birds' arrival in October to their departure in March. Each is identified, given a name, and its mate and number of cygnets recorded. Identification by means of the bill patterns has been supplemented by the use of large plastic rings, each individually numbered, and birds thus marked have been followed on migration across Europe with observers in many countries contributing to their sightings. Many have proved remarkably faithful to Slimbridge, returning for up to eleven successive winters, while others may desert the area for a year or two before returning once more. Cygnets accompany their parents throughout the first winter of their life, thereby learning about the migration route and wintering area, and then depart with them on spring migration. However before the breeding grounds are reached the young leave their parents to spend the summer with other immatures on some suitable wetland. They return on autumn migration separately but may link up with their parents again on arrival in the winter quarters. Family groups at Slimbridge have consisted of the parents, their current cygnets, and the young of the two previous years. Both types of swan pair in their second year of life, and breed when they are three or four. In the first year or two they will probably lay fewer eggs and rear fewer young than older, more experienced birds.

Bewick's Swans breed entirely within the Soviet Union, and so far no western biologist has been able to penetrate their breeding grounds to study them there. In arctic North America, however, there are no restrictions and work has now started on the Whistling Swan in the summer. So far it has been confirmed that, in common with many other bird species, a pair will return to the exact same nesting lake each year. Thus they use traditional localities at both ends of their long migration, and almost certainly use similarly well-known stop-over places on migration as well. In this way the hazards of life are reduced by the gradual acquisition of knowledge of the best feeding areas and the safest nesting sites and winter roosts.

Whooper Swan
Cygnus cygnus
The Whooper Swan nests throughout Iceland, which is only partly low arctic, while right across northern Europe and Asia it breeds within the boreal zone, or in a few areas just gets into the southern part of the tundra. Thus it only just qualifies as an arctic species. Its subspecific relative in North America, the Trumpeter Swan, does not penetrate the true arctic anywhere within its range, though it does occur in southern parts of Alaska, south of the tree line. Being rather larger birds than the Bewick's or Whistling Swans, the Whooper's young require at least another two weeks to fledge and this prevents it breeding further north, thus also avoiding competition with the Bewick's Swan. The Icelandic Whoopers have been categorised as a separate race by some authorities but the evidence is not conclusive. They do, however, lay smaller

Bewick's Swans alighting on a lake. Each winter they return to the same haunts.

clutches than the Whoopers of Scandinavia and Russia, and the eggs may be slightly smaller, too. The reason is that the Icelandic birds have a sea crossing of at least 500 miles to make from their winter quarters in Britain, and have no opportunity for replenishing their fat reserves, while those breeding in Scandinavia and Russia have an overland journey on which they can make frequent stops for rest and food.

There are probably about 5000 Whooper Swans in Iceland, of which the majority winter on freshwater lakes and floods in the British Isles, especially Scotland and Ireland, while the remainder, perhaps 500, stay in Iceland. They resort to the few ice-free waters, either on the coast, or inland where hot springs keep the water from freezing. Between 15,000 and 20,000 winter in the Baltic, particularly in the shallow area between Denmark and southern Sweden, and in the Netherlands, and these come from breeding areas in Scandinavia and Russia. Birds breeding further east in Russia travel south-eastwards to winter in Japan and China.

Grey Geese

The term 'grey geese' is given to those members of the genus *Anser* that are predominantly grey-brown in colour. Three species are arctic breeders while two more just reach the arctic but live mainly to the south, and although all five occur in Eurasia, only the Whitefront breeds in North America. The Bean and White-fronted Geese are additionally divided into a number of subspecies.

Geese are colonial nesters. Only the female incubates while the male stands nearby acting as sentinel, but once the young hatch both parents look after them, and the family stays together as a unit throughout the first autumn and winter, only splitting up when they return to the breeding grounds in the spring. There are several examples of moult migration undertaken by these immature geese. In winter, grey geese are almost always found in large flocks in traditional wintering areas, where they are frequently shot for sport and food. In summer, too, flightless flocks have been rounded up and slaughtered for food by northern tribes, a practice that has nearly died out. Nowadays flightless geese are more likely to be rounded up by biologists wishing to ring them and study their migrations and other habits. In winter they are caught for ringing in large rocket-propelled nets that are flung over feeding flocks.

Bean Goose
Anser fabalis
The Bean Goose breeds right across Europe and Asia from northern Scandinavia to eastern Siberia. It has evolved into a number of subspecies, only two of which are truly arctic, and the usual taxonomic treatment is to consider five subspecies of which *serroriostris* and *rossicus* breed in the tundra, while *fabalis, johanseni,* and *middendorfi* are confined to the wooded boreal zone to the south. The two arctic breeding subspecies are smaller than their relatives, thus contradicting Bergman's Rule which, as explained previously, says that the larger birds tend to nest further north. However, for a bird as big as a goose, obtaining enough food is perhaps of greater importance than the possibly marginal benefit of conserving a little more heat. Nevertheless it is of interest that the arctic Bean Geese have smaller bills and rounder bodies than their forest relatives, both adaptations to help conservation of heat.

The birds nest in colonies on the tundra, selecting dry hummocks, islands, and river banks as sites, and the nests are typical of geese, being low mounds of vegetation with a central depression. The female lays a clutch of from four to seven eggs, and then plucks down from her breast with which to line the nest. This provides thermal insulation as well as camouflage, as she covers the eggs with the down when she is off feeding, probably twice in each twenty-four hours (the male never sits on the eggs). The incubation period is four weeks. Families from a colony gather into a loose flock and feed on nearby marshland while the young are fledging—about seven weeks—after which the flocks begin to move south to their winter quarters. Here they live on farmland, mostly permanent pasture, but also on arable land, particularly stubbles in the autumn.

Bean Geese winter in northwest Europe, around the Mediterranean and Black Sea, and in China and Japan. They have been counted in some of their wintering areas but not accurately enough to be able to detect any trends. Formerly they were much more common in Britain than they are now but it is not possible to say how much this is an actual decrease in a population, and how much is due to a shift in range.

Pink-footed Goose
Anser brachyrhynchus
Pinkfeet breed in East Greenland, Iceland, and Spitsbergen. Large-scale ringing has shown that the Spitsbergen birds winter in the Netherlands. while the Greenland and Iceland birds winter solely within Scotland and Ireland. The two populations are therefore discrete both in summer and winter, but despite this there are no apparent detectable differences between them, suggesting that they have only separated a fairly short time in evolutionary terms. Their habits, too, are largely the same. The birds nest colonially, with the nest sites placed on low hummocks in the tundra, on rocky outcrops, and even on ledges and pinnacles in river gorges. Each site is chosen partly for its inaccessibility to foxes, partly for the provision of a good all-round view for the sitting bird and its mate, and partly, and this is true particularly in the more northern areas of the range, because such sites are free of snow earliest in the spring.

Pink-footed Geese migrate from breeding grounds in Greenland, Iceland and Spitsbergen to winter haunts in Britain and the Netherlands.

The Red-throated Diver [Red-throated Loon] usually breeds on small pools and flies to feed in larger lakes or the sea, bringing fish back for its young. Its streamlined shape is perfectly adapted to swimming and diving, but walking is very difficult. Opposite are two views of the species.

The largest single colony of Pinkfeet, in the middle of Iceland's upland lava desert, contains more pairs than all the rest of the breeding sites put together. It is in an oasis of marshland called Thjorsarver, and was first discovered in 1951 when Sir Peter Scott led a small expedition there. Previously only very small numbers of breeding Pinkfeet had been found in Iceland, many fewer than were necessary to produce the numbers wintering in Britain. Thjorsarver is very inaccessible, being reached overland only by a 50 mile pony trek or an even longer truck journey over stony tracks and across lava fields. Before 1951 the only known visitors were a small group of Icelandic farmers who penetrated there in late August each year, after the geese had finished breeding (their purpose was to round up the few hundred sheep that managed to reach the oasis earlier each summer). Scott returned there in 1953 and ringed about 10,000 birds, rounding them up before the young could fly and while the adults were flightless in their wing moult. His estimate of the number of pairs was around 3,500. The

geese were nesting on low banks, either side of the many streams and rivers running through the marsh, and quite thickly on frost-heaved tundra where there were numerous hummocks and ridges above the level reached by the melt-water as it thawed each spring. The birds clearly felt safe nesting on the ground with little or no protection from predators. Their only enemies were one or two families of arctic foxes, for whom the abundance of food in the summer was followed by an almost complete dearth in winter and this, coupled with the complete isolation of the oasis from other fertile areas of Iceland, prevented them from ever increasing above a minimal level.

In 1970 the author took part in a helicopter survey of the area to see how many pairs were then breeding there. In the seventeen years since the previous estimate the population, as measured by winter counts in Britain, had more than doubled to over 60,000, and plans to create a hydro-electric reservoir in the oasis made a fresh survey imperative. The total was found to be about 10,500 pairs, concentrated into about 32 square miles, making it one of the densest known colonies of geese in the arctic. The threat of a reservoir has receded but the opportunity was taken for some detailed biological studies. It has been found that at the present level the area is probably about at its peak capacity. There is no shortage of nesting sites but the necessary fertile marshland needed to provide food for so many pairs and their young probably cannot carry many more. It has also been discovered that some of the sites are as much as thirty or forty years old, and there is strong circumstantial evidence that pairs return to the same nest each year. The sites can be seen whether or not they are in use, as each has become built up into a permanent cup with a low rim comprised of compacted old droppings and nest material. Vegetation is often growing on these rims, and as it dies down each year it adds to the cup.

The larger of the two arctic swans is the Whooper. Its breeding season is too long to allow it to nest further north than the low arctic.

A striking feature of this vast colony is the absence of non-breeding birds, the one and two year olds not yet mature enough to breed. Observations in northern Iceland have revealed an extensive northward movement of birds during June and early July, and this is clearly made up of the immature geese heading on a moult migration to East Greenland. There are some hundreds of breeding Pinkfeet already there, but their numbers are completely swamped by the moult migrators from Iceland which may number 20,000 or more. In East Greenland they find large, relatively fertile areas of tundra, particularly in the bottoms of the great U-shaped glaciated valleys and on the rolling, heath-covered hills around Scoresby Sound. Both here and in Iceland the geese change their diet from grasses and sedges and eat the berries of the many fruiting dwarf shrubs as they ripen in the autumn. It seems that this diet, which contains a high sugar content, gives them a better energy value as they build up their reserves for the autumn migration south to Britain.

Currently there are about 90,000 Pinkfeet in the Iceland/Greenland population wintering in Britain, and a further 15,000 belonging to the Spitsbergen population. They are found in large flocks of several thousand, feeding almost exclusively on farmland. They start on the barley stubbles in the autumn, gleaning the spilt grain, move on to the potato fields as they are cleared, taking the waste and broken pieces, then feed on grass and growing winter wheat. Each dawn and dusk they fly to their feeding grounds, perhaps as much as twenty miles from their roost. The largest of these, always some safe loch or river estuary, can hold up to 25,000 geese.

White-fronted Goose
Anser albifrons
There are four or five races of White-fronted Goose, depending on who is one's favourite taxonomist. The one that breeds across

The White-fronted Goose's breeding and winter ranges, and its principal routes of migration.

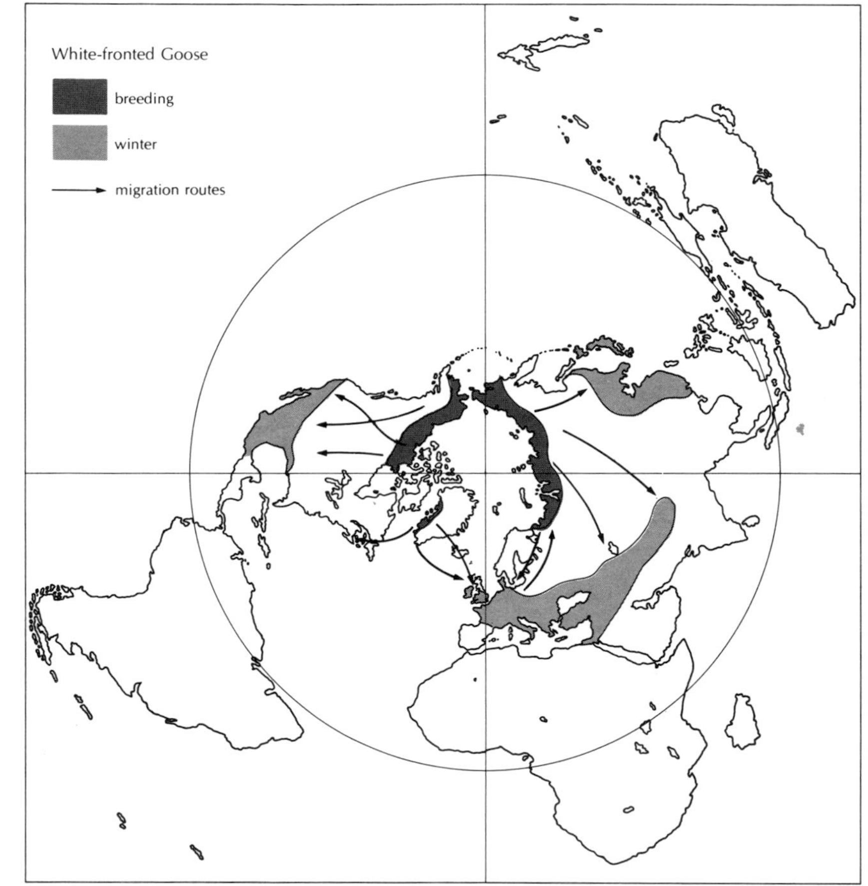

arctic Russia at least as far as the Kolyma River is the nominate *albifrons*. This winters in north-west, central, and southern Europe, northern India, China, and Japan. Counts and estimates make it by far the most numerous goose in Europe, with at least 500,000 birds. The Greenland Whitefront, *flavirostris*, breeds in western Greenland and migrates to winter in Ireland and western Britain, but its population is quite small at about 12,000. Its migration route takes it over the Greenland icecap and in the spring large numbers pass through Iceland. In North America there are two, possibly three, races. *Frontalis*, the Pacific Whitefront, is by far the commonest and most widespread, breeding from central northern Canada across Alaska and into eastern Siberia. The Tule Goose, *gambelli*, and the very recently named Elgas' Goose, *elgasi*, are apparently separable races but their exact breeding grounds have not yet been located. Their populations are probably quite small, and may one day be found in some area where confusion with *frontalis* has prevented their earlier discovery. They winter right down the western seaboard of the continent as far as Mexico.

Whitefronts nest in scattered colonies on the tundra, placing their nest on any hummock free of snow and melt-water. The nest itself is a shallow cup, often used year after year, and lined with bits of vegetation and the female's down. The usual clutch is four to six but the Greenland race tends to have up to eight eggs, and the incubation period is four weeks, with a further seven weeks for fledging. In common with the other geese, they are traditional birds, keeping together in family parties through the autumn migration and first winter so that the young birds learn by experience the best places to stop during migration, the route to take, and the best wintering grounds where food and safety are optimal. Such traditions help ensure the survival of the species.

In winter the birds prefer old pasture and wet meadows but have recently turned more to arable farmland. This is especially true of the Greenland race which was once confined to the acid bogs of Ireland and Scotland. They moved nto a large area of permanent, wet pasture on the Wexford Slobs in south-east Ireland before the war, and gradually built up until about one-third to one-half of the population was concentrated in this one area. Then it was proposed to plough the grass and grow crops. After much protest from conservationists a small area of grassland was set aside for the geese, but they quickly adapted to feeding on stubble and potato fields like other grey geese, and the grassland is now hardly touched. Although most geese, including the majority of Whitefronts, winter in large flocks of many thousands, the Greenland race is distinctive in occurring in only small numbers in some areas, often no more than fifteen or twenty birds.

Lesser White-fronted Goose
Anser erythropus
This species bears the same geographical relationship and therefore habitat relationship to the Whitefront as the Whooper Swan does to the Bewick's. In other words it nests just to the south of the Whitefront, occupying the forested and scrub zones bordering the tundra, which it only penetrates in a few areas. However in contrast, here Bergman's rule is followed, with the smaller of the pair breeding to the south of the larger. The Lesser Whitefront nests from Scandinavia, where it has decreased in recent years, through northern Russia to eastern Siberia. Most of the western birds migrate to the shores of the Black and Caspian Seas, but stragglers turn up each year in western Europe, invariably with flocks of Whitefronts or Bean Geese. They presumably meet up with these species on migration, and the occasional 'lost' birds tag on to the flocks of other geese out of a sense of gregariousness—any goose flock is a safer place to be than on one's own.

The spring thaw comes to the Alaskan tundra. This pair of White-fronted Geese will nest on a dry ridge protected by the water.

The V-formation adopted by geese in flight, including these Greylags, is the best for maintaining visual contact.

Greylag Goose
Anser anser
The only area where the Greylag lays any claim to be arctic is in Iceland. Here about 70,000 birds breed. Whereas the Pinkfeet breed in the centre of the country, confined to the one major colony in the Thjorsarver oasis plus much smaller numbers scattered along river gorges, the Greylags occur right round the coastal belt, and also in the broad glacial valleys of the north of the country. Unlike all other goose species, they are not easy to catch when flightless. Instead of bunching and allowing themselves to be driven into a pen, the birds commonly scatter and dive, refusing to be herded in any direction. Perhaps it is this ability that has enabled them to survive the pressures of living near man, for they are undoubtedly good to eat and therefore a potentially valuable source of food, particularly in time of hardship. But in more prosperous times they are not loved, for they are accused of eating grass and crops. Icelandic Greylags winter exclusively in the British Isles.

Snow Geese

There are three arctic snow geese, the Greater and Lesser, which are subspecies, and the Ross's Goose. The Emperor, although not pure white, is usually classed with the snow geese. The Lesser Snow and the Emperor breed in both North America and eastern Siberia and are good examples of species that, having survived the ice age in Beringia, continue to live on both sides of the Bering Straits after the disappearance of the land bridge. Further evidence of a link between the two areas is that both birds whether breeding in Siberia or Alaska, migrate to the western coasts of North America for the winter. Snow Geese en masse produce possibly the finest of all wildfowl spectacles. Their black and white plumage seen against a blue sky and green fields coupled with the clamouring of their voices can hardly be equalled for the thrill and pleasure it brings to the watcher.

Lesser Snow Goose and **Blue Goose**
Anser caerulescens caerulescens
The Lesser Snow Goose has two colour phases, the white and the blue. The latter is frequently called the Blue Goose, and it is most common at the eastern end of the breeding range in arctic Canada, but is steadily increasing in the western areas too. Hybrid pairs between the phases are common. Lesser Snow and Blue Geese nest in large colonies on the tundra, their main breeding habits being much as other geese, though their incubation period is rather shorter than most of the grey geese at 22 to 23 days. A detailed study of them in the McConnell River area of the North West Territories revealed much about the stresses experienced by geese breeding in the arctic, and it seems probable that what was found for them will be true for other species as well. One particular interest of the study was to discover how many eggs were lost between laying and hatching, and from what major causes. It was found that about 20% of eggs were taken, mostly fairly late in incubation. Arctic Skuas (Parasitic Jaegers) and Herring Gulls, both very adept at finding eggs, were the main predators, being attracted to the colony for its

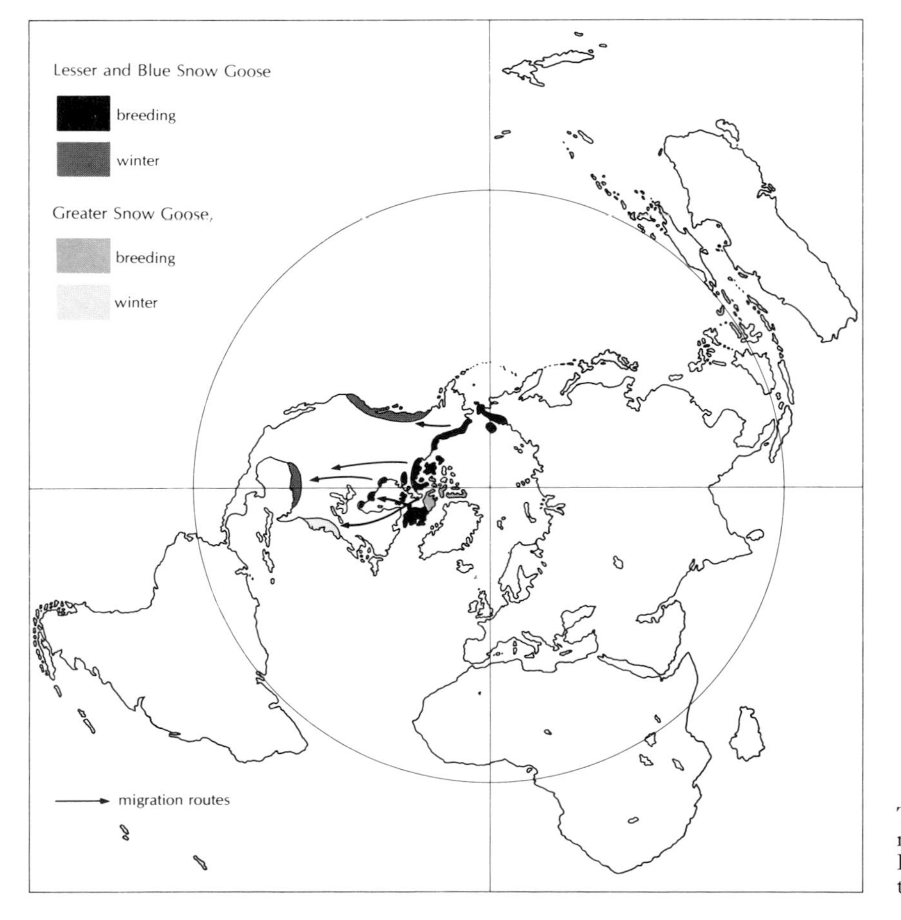

The breeding and winter ranges of the Lesser and Blue Snow Goose, and the Greater Snow Goose.

potentially rich source of food. However the incubating geese and their mates were equally good at defending their nests strongly against these avian thieves, and eggs were only being taken from nests which the parents had left, if only for a short time. Investigating why the birds should leave their nests disclosed that many were becoming so short of food late on in incubation that they were being forced to spend longer and longer periods off feeding.

Looking further back along the chain of events, it is known that whereas an incubating female might normally expect to lose about 25% of her peak spring weight by the time the eggs hatch, adverse weather conditions can cause this to increase sharply. The loss of weight by a bird is directly linked to its loss of heat, which in turn is governed by the prevailing weather. Spells of cold winds, coupled with low air temperatures, can therefore greatly increase weight loss unless the bird is able to replenish its energy by feeding more. The presence of predators dictates that those birds which leave their eggs to feed in order to stay alive themselves, lose their eggs instead. However, in the study it was found that some incubating females stayed on the nest too long, losing so much weight that they eventually died. No less than 140 birds were found dead on their nests, mostly in a sleeping position with the head tucked under the wing. They were all extremely thin, with an average of 43% (less than half) of their spring weight. The strength of their incubating instinct had carried them past the point at which it would have been prudent for them to desert the eggs and go and feed.

The Lesser and Blue Snow Geese are probably the most numerous goose species in the world, with winter counts in North America totalling over one and a half million. The birds winter practically all round the continent, from California to Mexico, and along the Gulf coast to Florida, and from South Carolina to New Jersey. They live on coastal lagoons, marshes and floods, feeding exclusively on farmland.

Greater Snow Goose

Anser caerulescens atlantica

In contrast to the Lesser and Blue Geese which are low arctic, this is an exclusively high arctic breeder, demonstrated in its larger size (thus confirming Bergman's Rule for a change) and in its lack of a colour phase,

Greater Snow Geese on their wintering grounds in the eastern United States. They feed on the roots and stems of marsh plants.

having just the single pure white plumage. The nesting area is in the north-eastern islands of the Canadian arctic, Ellesmere, Axel Heiberg, Prince of Wales , and so on. From these islands it migrates south through Canada with a very regular stop-over place, in both spring and autumn, at Cap Tourmente on the St Lawrence River in Quebec Province. Counts have been made here since the turn of the century and they show a steady increase in numbers, which in recent years has escalated remarkably fast. In about 1900 there were no more than 3,000 birds. They increased slowly to 6,000 by 1921 and to 10,000 by 1937. This figure was doubled by 1941, was up further to 30,000 by 1951, and in 1958 there were nearly 50,000. After levelling off here for a few years they increased very sharply to 80,000 in 1971 and to no less than 200,000 in 1975.

The slow but steady increase in the first half of the century may have been due to better conservation as more knowledge of the population led to controls on the former indiscriminate shooting. The staggering increase of recent years was the direct result of three good breeding seasons with almost all the potential parents raising large families of young. Unfortunately, these huge numbers have now brought the species into conflict with agricultural interests, particularly in the Cap Tourmenté area where the *Scirpus* marshes that used to be the sole feeding area for the geese cannot support the much larger flocks. The birds have now moved inland to some of the adjacent farmland where they graze on growing crops. What to do about an arctic breeding species that is subject to great fluctuations in population size because of the rigours of the climate in its breeding area, is one of the main conservation dilemmas facing waterfowl biologists. Recent good breeding conditions have produced very healthy populations of many arctic geese, but several of these live on farmland during the winter. Where small numbers might be tolerated, large numbers bring increasing complaints of damage to crops. Any control measures instituted must be set against the background of a possible decline in numbers from natural causes due to bad arctic summers—over-doing it could reduce a population to a level from which it might have great difficulty in recovering.

Ross's Goose
Anser rossicus
It was only in 1945 that the breeding grounds of this little snow goose were discovered, in the North West Territories on the mainland coast of arctic Canada. Even now the full extent of its breeding range is imperfectly known. Thirty years ago it was regarded as being very rare with a population of only 3,000 individuals. Gradually however it became apparent that it was in fact much more numerous, for it was realised that its wintering flocks had previously been wrongly identified as Lesser Snow Geese. When the confusion was sorted out, flocks were found wintering in much larger numbers and over a much wider area than earlier suspected. From 3,000 the total leapt to 25,000, then to 40,000, and even the present day estimate of 70,000 may be on the low side.

As previously stated, Ross's Geese use up no less than 86% of the frost-free period for their breeding cycle. This near maximum use of the potential time means that a late spring could inhibit breeding altogether. They nest almost exclusively on islands in lakes on the Canadian tundra. The islands may be only a few feet across or hundreds of yards, but their one similarity is that early meltwater prevents arctic foxes from getting to them once the geese have started to breed. Here they are safe from the one ground predator that could damage their breeding prospects. Their principal wintering areas are on farmland in central California.

Emperor Goose
Anser canagicus
The range of this bird is very restricted, being only northeast Siberia, the coast of northwest Alaska, and on St Lawrence Island, lying in

between. In winter it does not move far south, staying for the most part in the Aleutian islands, but it also occurs down the Pacific coast of North America to British Columbia and California. It nests among rather wet tundra around small pools, never far from the coast, and immediately before and after the breeding season it resorts to the salt marshes right on the coast, feeding round the many tidal lagoons. The nests can be as close as twenty or thirty yards from each other, although it is not strictly a colonial species. Mostly, however, they tend to be hundreds of yards apart. Those examined in northeast Siberia were found to contain less down than nearby White-fronted Goose nests, and the eggs and young suffered greater losses from wet and cold than those of the latter species. This suggests that the Emperor Goose is less well adapted to the arctic than the Whitefront, and that it has probably reached the limit of its northward penetration.

As soon as her newly hatched brood are dry, the female Emperor Goose will lead them off the nest to feed.

Black Geese

As the name implies, these geese are generally black or very dark brown, though often with much grey and white. They are mostly smaller than the grey geese, though the Canada Goose does have some subspecies which are larger. Among their number are the Brent which, of all the geese, nest the furthest north and are consequently subject to the most extreme conditions. The general breeding habits of the black species are similar to the other geese.

Canada Goose
Branta canadensis

Although once an exclusively North American species, the Canada Goose has been introduced by man into Britain, Sweden, and New Zealand. In North America it has evolved into a large number of subspecies. One observer, a confirmed 'splitter,' is nearing as many as thirty, but eleven is the generally accepted number. However even eleven is probably too many as recent knowledge has shown that some of these intergrade

This Canada Goose has just finished laying her clutch of eggs and has yet to add the insulating and concealing down.

over a wide area so that experienced observers are unable to name the races that they watch in the field, and even museum men have great difficulty identifying them.

There is considerable difference in size between the various races of the Canada Goose, with the largest, *maxima,* four times heavier than the smallest, *minima.* The four races that breed in the arctic, *minima, taverneri, parvipes,* and *hutchinsii,* are all smaller than the remaining races which breed to the south. At first sight this is another example of geese, as large birds, disobeying Bergman's Rule but, on the other hand, it has been shown that all four of these small races winter to the south of the larger ones. With the exception of the Whitefront and the Greater Snow, the majority of the swans and geese appear to obey Bergman's Rule only in the winter and not in the summer. Quite why there should be this difference is not at all clear. There is an obvious advantage in being larger when breeding or wintering in a cool or cold area but equally there is an advantage in not being too large when living in an area where the food supply is likely to be limited. Each species has apparently worked out its own solution, but it is interesting that different birds should have arrived at exactly opposite answers.

Some comparative figures show that over a very wide area of tundra, Canada Geese complete their egg laying within a 10-day period in the spring, while in California, with its much milder climate, laying by the same species is spread over as much as 50 days. Similarly the incubation period in the

The ranges of breeding and wintering of the Barnacle Goose, and its main migration routes.

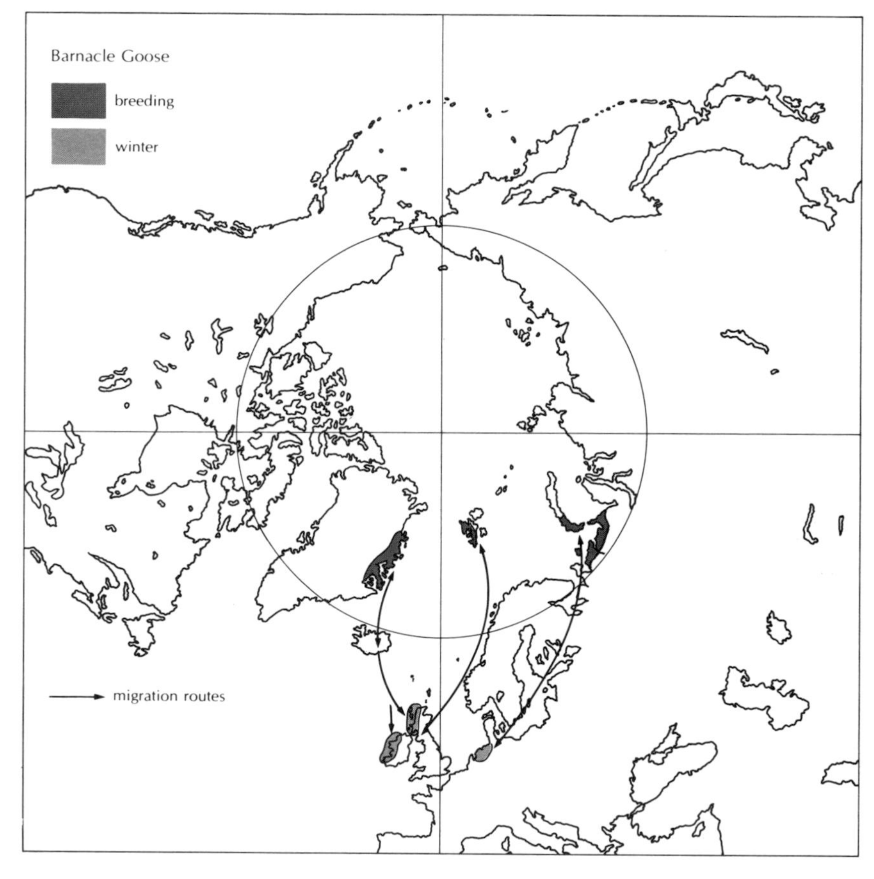

Opposite: A flock of Barnacle Geese on a Hebridean island. They are grazing birds taking leaves and stems of grass and other pasture plants.

arctic is 24 to 25 days compared to 28 to 32 days, and the fledging period of the young and the moult period of the adults are also less. All of these are vital adaptations to the very short northern summer and all will be helped by the birds being small rather than large in size.

These birds are not as colonial as most other geese. They breed in widely scattered fashion over the tundra, with a density of no more than a few pairs per square mile. The principal value of a colony is that it provides protection in numbers against predators, which while they may take the eggs or young of some birds will never be numerous enough to seriously affect the success of the colony as a whole. This defence may not be much greater than that offered by nesting in a very dispersed manner, for here a predator will have to rely much more on chance in finding a nest or family of young, and so losses will still not be too high. Scattered nesting also allows the birds to exploit very small areas of suitable feeding that could not support more than the single family. One measurement of the losses of eggs to predators showed that Arctic Skuas [Parasitic Jaegers] were the main culprits, taking in an average year 10% of eggs laid—only half the loss rate reported for colonially nesting Snow Geese. Furthermore in the case of the Canada Geese it was found that in years with high lemming numbers, a cycle that occurs every four to seven years, egg losses were much reduced as the skuas had such a plentiful supply of an alternative food.

Barnacle Goose
Branta leucopsis
The three separate populations of this species breed in East Greenland, Spitsbergen, and northwest arctic Russia. Their wintering grounds are also discrete, being respectively the western islands of Scotland and Ireland, the Solway Firth on the borders of Scotland and England, and the Netherlands. Although some of the wintering grounds of the different groups are only 100 miles apart, ringing (banding) shows that there is no mixing of the three stocks.

The first Barnacle Geese to be found breeding were on ledges on cliffs along the sides of valleys, up to 300 feet above the ground. This is still the usual site in East Greenland, but the Spitsbergen birds have shown an interesting change in preference in the last few decades. They have turned much more to small off-shore islands, which may be as little as twenty yards across or up to several hundred acres in extent. The islands themselves often cannot support the families of growing young, and the parents therefore swim their brood across the sea, up to two miles or more, to the mainland and nearest marshy area. In one area, three offshore islands support a breeding population of between 60 and 80 pairs, while up to 500 non-breeding birds may also be present during the moult period. The mainland coast opposite has only a narrow coastal strip of flat ground, hemmed in by steeply rising scree slopes running up to a rocky wall with peaks about 2,000 feet high. These slopes contain an enormous Little Auk colony that runs more or less continuously for about ten miles, and may contain anything up to four million or more birds. Here a unique food chain takes place. The auks feed mainly on plankton and small fish from the rich polar seas. Their staggering numbers then bring to land vast quantities of nutriment in the form of their droppings, which are full of nitrogen and other minerals. This richness gets washed out of the scree slopes by the melting snow and in turn produces a comparatively lush and fertile marsh on the coastal plain, providing ample feeding for the Barnacle Geese.

The island nesting is of course to escape from arctic foxes, and this too is the reason for the cliff sites used in Greenland. Here small colonies occur whose size is probably limited by the number of available ledges. The average ledge used is about two yards wide and up to two and a half feet deep. Anything

much smaller and the bird will have difficulty in landing. This safe positioning of the nest, a couple of hundred feet above the ground and perhaps with a steep scree slope beneath it, poses something of a problem for the gosling when it has hatched and is ready to move to the feeding marshes below. There are two published accounts by apparently reputable observers alleging that the young are carried down to the ground by their parents, either riding on their backs, or perhaps being carried in their bill. Such behaviour is totally outside the normal range of actions of a goose, however, and to many people seems highly unlikely. The alternative method is for the goslings to fall, tumble, scramble, and generally bounce their way down. No cliff is so sheer that there is a vertical drop free of all ledges and obstructions right to the ground, and a fall of even fifty feet by a light downy bird is not going to hurt it in the least. Canada Goose young have been watched falling from this height from a relatively low cliff-ledge nest, and the young were perfectly all right afterwards apart from the apparent slight winding of one or two. It is an interesting problem, but clearly not insuperable, otherwise the geese would hardly continue to nest on such cliffs.

Brent Goose

Branta bernicla

One of the most northerly breeding geese, the Brent occurs in the Canadian arctic islands, northern Greenland, Spitsbergen, Franz Josef Land, and northern Siberia. There are three races but two of these intergrade where they meet in arctic Canada.

Nesting so far north they meet more extreme conditions than any other goose and consequently suffer from a higher proportion of breeding failures. The success of a population is measured by biologists from counts of the proportion of young birds and the mean brood size when the geese reach their winter quarters. An average season will produce a proportion of between 20% and 30% of young birds with a mean brood size of 2·5 to 3·0. A really good season when virtually every pair capable of breeding has done so, and done so well, will raise the percentage of young to as high as 50%, and the brood size may be 3·5 to 4·0. But in some years it is possible to scan through several thousand birds without seeing a single young one. Such complete failures stem principally from a very late, cold spring where the thaw is delayed and where there may even be further snowfalls. This combination leaves nowhere for the geese to nest, all possible sites being under snow, and the prevailing weather conditions anything but propitious for rearing young. It appears that the birds are capable of moving on to look for a better area, but clearly they are going to have to find somewhere quite quickly, just as they cannot afford to wait for long at the original site.

If the ground is snow covered not only will nest sites be hidden, but there will be no food available for the geese either. Their energy reserves become depleted and after a certain point the egg follicles of the females are absorbed back into the body in an effort to retain this valuable form of nourishment. It is not certain whether this follicular atresia, as it is called, can be arrested once begun or even go into reverse should conditions become more favourable, but if the snow persists too long all the follicles disappear and any chance of breeding is over for that year. The atresia not only provides nourishment for female goose at a critical time but also serves to prevent very late nesting, which would put the whole population at risk by delaying the completion of the breeding and moulting cycles beyond the end of the normal frost-free summer.

The years of complete non-breeding occur with some regularity, but rarely more often than one year in three. Clearly it would only take some very marginal climatic change to increase it to a frequency that might seriously affect the future of the population. In fact quite the opposite has occurred with the population of Dark-bellied Brent Geese which breeds in Siberia and winters in northwest

Europe, mainly Britain and France. This population numbered 25,000 to 30,000 in 1967 but increased to nearly 90,000 in the next six years, thanks to a succession of good breeding seasons. Between 1969 and 1973 only 1971 was a bad year, all the others being well above average. Although this has brought a comparatively scarce species up to what might seem a very safe level, it must be remembered that what goes up can equally well come down; that a run of good years may well be followed by a run of bad, and all of the increase lost very quickly.

One population of Light-bellied Brent winters in Ireland. It was long thought that these geese bred in northern Greenland, where the species is known to occur. However the fairly small numbers of the bird up there, not to mention the paucity of available habitat, led people to suppose that some at least might breed in the northeast of arctic Canada, particularly on Ellesmere Island. This supposition was neatly confirmed in 1970 when two geese marked on Ellesmere were shot in Ireland. Since then some further work in arctic Canada in summer 1974 has pushed the known breeding area much further west, with birds marked on Bathurst Island (100°W) being reported in Ireland. Exactly which route they take on their long migration is still a matter of doubt. It is known that perhaps half or more pass through Iceland in both spring and summer, and it must be supposed that they either fly right round the northern

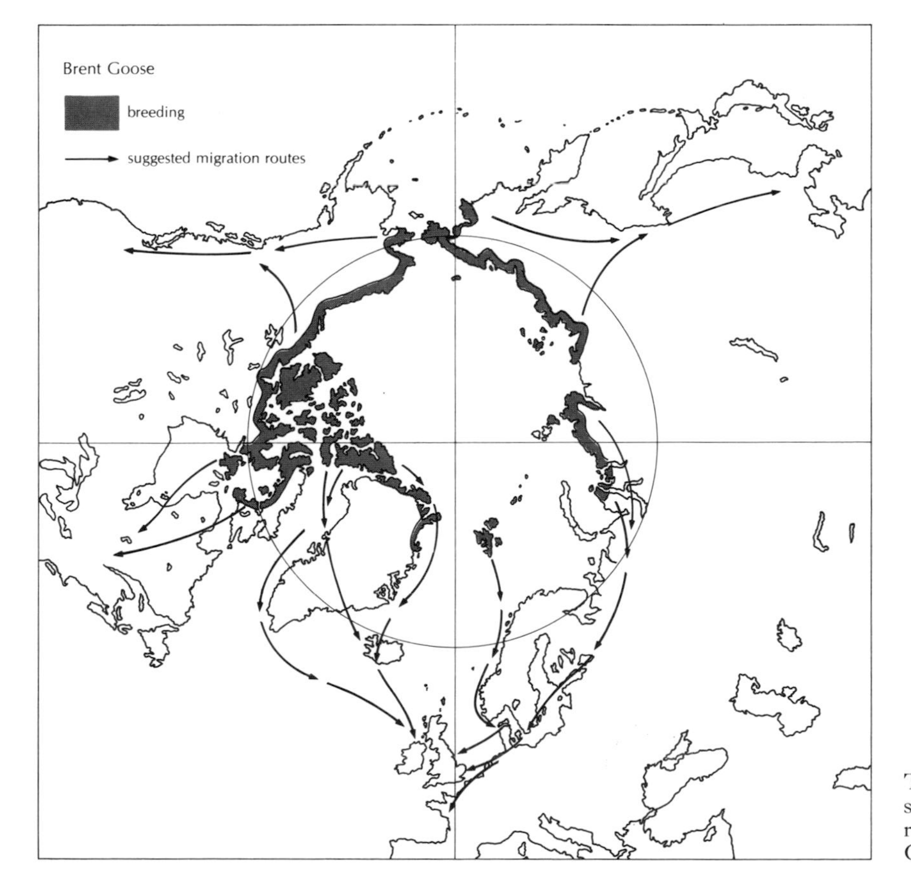

The breeding areas and suggested migration routes of the Brent Goose.

tip of Greenland or more probably straight across the icecap by the shortest route. Other Brent are known to migrate down the west coast of Greenland in autumn and these may make their crossing to Ireland from the southern tip of the country, at Cape Farewell, thus avoiding the icecap crossing but giving themselves a far longer oversea journey instead.

There are other Light-bellied Brent breeding in arctic Canada that winter on the Atlantic coast of North America, and the separation of these from the Irish birds in the summer has not yet been worked out. Further west, the Pacific or Black Brent breeds in western arctic Canada, northern Alaska, and eastern Siberia, and winters on the Pacific coast of America. The winter habitat of all the Brent is estuaries and tidal mudflats.

Red-breasted Goose
Branta ruficollis

Twenty years ago there were probably over 30,000 Redbreasts alive, but the total now may well not be more than half that figure, making this attractive small goose one of the most threatened of waterfowl species. The reasons are complex and include changes on their wintering grounds, which themselves have shifted in the last few decades from near the Caspian Sea to Romania, and perhaps climatic changes and a succession of cold winters. Their breeding range in the central Siberian arctic seems safe enough but even here there may be adverse factors.

The bird has not been studied much in the arctic but the most recent information was gathered by two Russian biologists during the 1960s, who reported one of the most extraordinary avian relationships to occur, certainly in the arctic, and possible anywhere. They found over 100 Red-breasted Goose nests with almost every one close to a Peregrine's nest. The few that were not, were among Herring or Glaucous Gull's nests or beside that of a Rough-legged Buzzard. There were up to five nests beside each Peregrine, always within 300 feet and usually no more than 30 to 100 feet, and one goose was sitting on its eggs only five feet away. The groups of immature non-breeders that occur in any population of geese also tended to be found in the same vicinity.

Of the 22 Peregrine nests the Russians found, most of them quite far apart with from four to six miles between them, 19 had geese nesting nearby. The relationship between the falcon and the geese is thus very close indeed, and so far no-one has found a Red-breasted Goose's nest other than near that of a bird of prey or, very occasionally, a large gull. There was evidence from old and unused nests that the geese were formerly more numerous. Instead of the present five, as many as 20 or 30 may once have formed each colony. The reason for this decline is not certain, but it could be related to changes on the wintering grounds. However a further threat hangs over them: Peregrine populations in many parts of the world have declined through the birds absorbing toxic levels of pesticides from their prey. It is now thought that the Siberian Peregrines are now affected, and this may be having a serious effect on the Redbreasts.

The protection afforded to the Redbreasts must be against the arctic fox. The tundras of Siberia are mostly very flat, with few lake islands and no steep cliffs on which the birds might seek security. The falcons nest on such rising ground as there is, usually hummocky knolls 50 or 100 feet above the general ground level, and so the geese nest there too, above any possible spring flood level. The Peregrines effectively keep the foxes from approaching too near their own nests and thus unwittingly defend the geese as well. There is no evidence that they ever prey on the geese, though in theory this could happen. They usually take flying prey so birds on the ground or water should normally be safe. Once the goslings hatch they will be taken by their parents to the nearest grazing marsh, with the safe haven of water not too far away.

Dabbling Ducks

There are six species of dabbling ducks which penetrate into the arctic, but only two qualify for full arctic status, the Mallard and the Pintail, and these only just. Dabbling ducks nest on the ground, usually in thick cover, though in the arctic they may have to be more in the open. The female selects the nest and lays a clutch of from eight to twelve or fourteen eggs. She then plucks quantities of down from her breast with which to line the nest to provide insulation for the eggs against damp and cold. The male takes no part in the incubation, and indeed deserts the female as soon as she has completed the clutch and begun sitting. The female therefore has to leave the nest about twice each day, to wash and feed, and when she does she pulls the down over the eggs, both to keep them warm and to hide them from predators.

Mallard

Anas platyrhynchos

The Mallard is the most widespread and adaptable of all wildfowl. Man has domesticated it, introduced it to new lands, and harvested it for centuries. In all the new countries to which it has been brought it has settled down and spread rapidly. Although primarily a species of temperate lands it is perhaps not surprising that it has spread north through the boreal zone to the edge of the arctic tundra. Occasional individuals and pairs turn up far to the north of the normal range and may, in a good season, rear a brood. Perhaps it was adventitious birds such as these that first reached southwest Greenland a few thousand years ago. Here they colonised the coastal belt of habitable land fringing the icecap, from Cape Farewell northwards up both coasts to about 66°N on the east coast and about 69°N on the milder west. Gradually the birds evolved into a subspecies distinct from the Mallard of North America and Eurasia. They became more or less sedentary, moving only short distances, and they also grew in size, obeying Bergman's Rule, so that they are now about 10% bigger than the normal form, although their bill is rather smaller.

Another obvious characteristic of the Greenland Mallard is the enlarged nasal glands which show as a prominent ridge on

A pair of Mallard jumping from the water. These are the most widespread of the ducks, reaching the low arctic in several areas.

the upper part of the bill. These glands are used to remove salt from the blood stream and are an indication of the largely maritime life of the birds, which only use freshwater for the short breeding season. Then they go inland to nest beside small pools and streams. In the autumn they move to the fjords, and when these freeze, out to the coast, where they keep to areas of open water in which currents or tidal movements prevent freezing. They feed on seaweed and other shoreline algae from rocks exposed at low tide, and also take molluscs and crustaceans, but severe conditions can cover even these areas with ice and therefore bring starvation and death to many. Yet another adaptation to life in the arctic has already been mentioned, namely the delaying of breeding until the birds are two years old compared with the usual one year old elsewhere. This means that they are more experienced when they do breed and are therefore more likely to rear young successfully in marginal conditions. Even so, the conditions in spring do not allow the female to produce many eggs—the normal clutch size is only seven to ten compared with nine to thirteen elsewhere.

Pintail

Anas acuta

The Eurasian range of the Pintail has long included the tundra area north of the boreal zone which is its true home, but it is a comparatively recent coloniser of several parts of

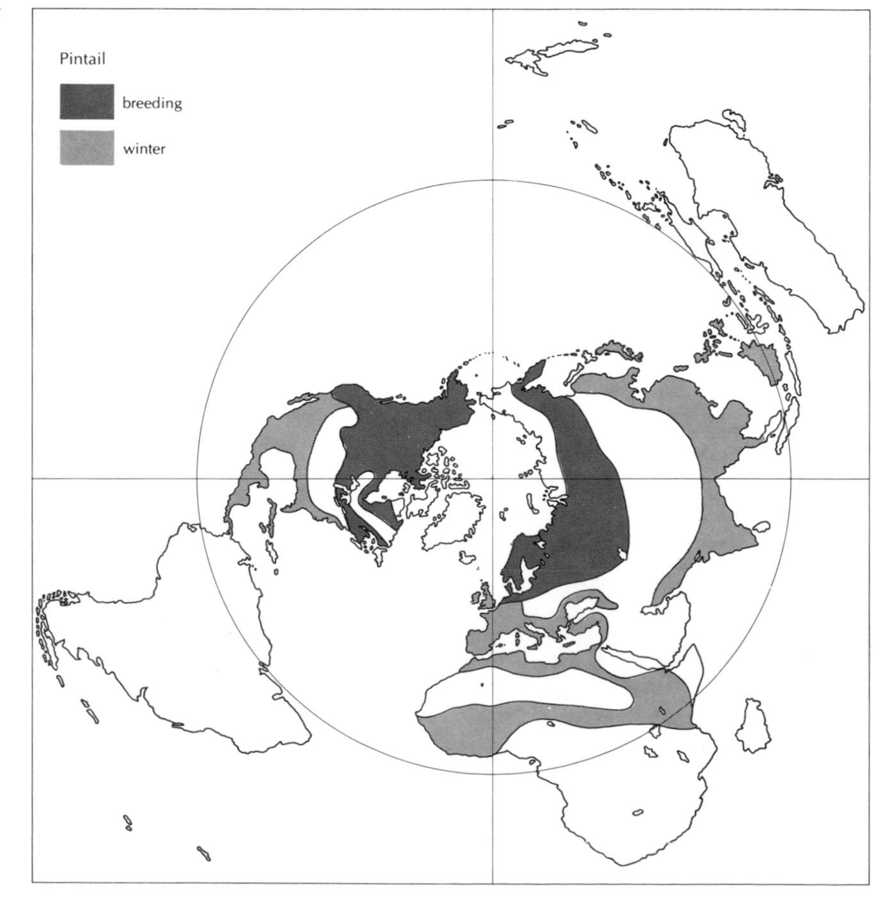

The Pintail's breeding and winter ranges.

Opposite: There are up to one and a half million Lesser Snow Geese in North America. They nest in large colonies, keeping in vast flocks on migration and during the winter.

A beautifully camouflaged female Pintail returning to her equally well concealed nest.

Opposite: The Brent Goose [Black Brant] of Alaska nests in low-lying tundra close to the coast. Its annual breeding success varies widely according to local weather conditions.

the arctic. There has been a steady spread north in eastern arctic Canada in the last thirty years, with breeding in Baffin Island confirmed in the 1940s, and a pair breeding on Ellesmere Island in latitude 82°N in 1967, though so far this has not been repeated. During the same period the species reached western Greenland, with breeding first reported from several localities in 1947 and 1948. Here conditions are eminently suitable for a migrant dabbling duck, and the Pintail has successfully filled the available niche.

They are indeed great migrants, making long journeys in both the Old and New worlds. Ringing has shown that there are quite complex movements in Europe, the full ramifications of which are not yet unravelled. It is known that those that winter in the British Isles mostly breed in the Soviet arctic and boreal zones to about 100°E. But some of the birds only use Britain as a migration stopping place before travelling on further south—so far there have been three recoveries in West Africa indicating a much longer movement than had previously been suspected. One bird was ringed in Suffolk in September and shot within a few weeks in Sénégal, close to the equator, and it is highly probable that it had already flown from Russia to England in the weeks immediately prior to being ringed.

The return route in spring is a different one, taking the birds further east through the Mediterranean to Italy and Turkey, before they head north through the Soviet Union to their breeding grounds. Thus in a year they will complete a very large, circular migration. It is perhaps not surprising that such long distance travellers should have spread north into the arctic, and that this spread is continuing. In winter, Pintails are primarily salt water birds, living in estuaries and the mouths of large rivers. In some areas however they also occur plentifully on freshwater floods.

Wigeon are highly gregarious ducks, flying in dense flocks, and feeding in winter in tightly packed groups on wet grassland.

Wigeon
Anas penelope
American Wigeon [Baldpate]
Anas americana
These two species inhabit the Old and New Worlds respectively, and fill approximately the same niche in each region, both just reaching the tundra zone. Their habits are similar in many respects. In winter they are of interest for their occurrence in very large flocks and for their main feeding behaviour of grazing on short grass marshes and fields. This is particularly pronounced in the European Wigeon which can be found in flocks of tens of thousands grazing on saltmarshes and other areas where the vegetation is kept down by cattle and sheep. Except in a few localities they are maritime species, roosting in estuaries and shallow coastal waters and grazing on the saltings.

Their nest is a small depression in the ground, lined with the available grasses and leaves. They are not strictly colonial but in favoured areas there may be many nests with only a few yards between them, and this is especially true on small islands where the birds concentrate to be free of foxes and other ground predators. The nest is usually placed in thick vegetation in such a way as to give some protection from overhead for the sitting female.

Teal
Anas crecca
Green-winged Teal
Anas crecca carolinensis
The smallest of the dabbling ducks, the Teal has successfully spread north to breed in the edge of the low arctic in both Canada and Eurasia. Being small its incubation period of 21 to 23 days is shorter than its relatives, as is the fledging period of four to five weeks compared with five to six weeks for, say, the Wigeon. This vital difference has enabled the occasional pair to breed well to the north of the usual range, including Spitsbergen at about 78°N. Like other freshwater dabbling ducks, the Teal is dependent on water for the rearing of its young, but may place the nest up to half a mile or more away from the nearest pool. It is also always well concealed and these two factors make it one of the

A flock of Teal led by a Spotted Redshank. Both species breed in the low arctic of north-west Russia.

hardest waterfowl nests to find. When the ducklings hatch the female has to lead them on the hazardous journey to the water. In winter the birds flock on freshwater marshes, and also on river estuaries and coastal lagoons. They migrate from their northerly breeding grounds for the winter but are not such long distance travellers as some of the other dabbling ducks.

Baikal Teal
Anas formosa
The range of this most attractive small duck, one of the less well known species, includes the great Lake Baikal in central Siberia from which it gets its name (the '*formosa*' of its Latin name means 'beautiful' and has nothing to do with the offshore Chinese island). Its breeding range extends northwards into the tundra of northeast Siberia but little is known of its breeding habits.

Diving Ducks, Sea Ducks, and Sawbills

Although a somewhat amorphous group, all its species resort to the sea at least in winter. Many of them breed beside it as well and all are principally animal feeders, getting their

food underwater. The normal clutch size is below that of dabbling ducks, though their general breeding habits are the same. The female chooses the nest site and prepares the nest cup, and she alone incubates the eggs and rears the brood. The sites are varied, with at least two species using holes in the ground, while the others either seek thick vegetation or conceal their nests among rocks. Some of the species are highly migratory, others move only as far as weather and food conditions dictate.

Scaup [**Greater Scaup**]
Aythya marila
The Scaup is not restricted to the arctic but nowhere occurs much to the south of 60°N. It is common in arctic Alaska, around Hudson Bay, in Iceland, northern Scandinavia, and across arctic Russia. The birds breed in colonies in some areas, with their nests only a few feet apart. These are on the ground, usually with the cover of either vegetation or a convenient boulder. The usual clutch is eight to ten eggs, and the female sits for 25 to 27 days. As with all ducks the young feed themselves and can dive within hours of leaving the nest. The adults feed mostly on molluscs, particularly bottom shellfish and other aquatic animals, while the ducklings take insects and their larvae.

In winter Scaup gather in very large flocks in favoured areas. These are mainly on salt water in Europe but the species is more a freshwater duck in North America. There are flocks totalling many tens of thousands in the shallow Baltic Sea, particularly in the area between Denmark and Sweden, while the largest concentration in Britain is in the Firth of Forth near Edinburgh. Here sewage outfalls and other wastes pour into the estuary, producing rich feeding conditions not only for up to 30,000 Scaup, but for several thousand Eiders, Goldeneyes, and Pochard as well. Extensive mussel beds have grown up in the shallows by the outfalls, their flesh too contaminated for human consumption but apparently not harmful to the ducks.

A half-grown Scaup rests on a rock with its mother. She will desert it as soon as it can fly.

Harlequin Duck

Histrionicus histrionicus

This attractive duck has a widespread range, broken into a number of discrete areas. One population, sometimes given a subspecific name of Pacific Harlequin, breeds in north-eastern Siberia, western Alaska, British Columbia, and the mountainous parts of California. In the east, they breed in southern Baffin Island, Labrador and parts of Quebec, western and southwestern Greenland, and Iceland.

The Harlequin has its own restricted ecological niche that is very obviously distinct from other ducks. It nests beside fast-flowing streams in which it is completely at home, swimming in the most turbulent rapids and feeding in the shallows by turning over the stones on the bottom. Here it catches fish fry and insect larvae of many kinds. Like other ducks living in similar habitat in other parts of the world (Salvadori's Duck of New Guinea and the Torrent Ducks of South America), it has a boldly patterned plumage including white patches and stripes which, although it looks quite conspicuous when the bird is frozen by the camera or paintbrush, in fact produces a fine camouflage effect against the light and dark of rushing water. Another characteristic of the Harlequin, and the other species, is the constant bobbing of its head backwards and forwards, an action that enables it to spot objects more readily in its constantly moving environment. Also the thigh and leg muscles of the adults and ducklings are larger and stronger than in other ducks that only have to swim in still water.

Their winter habitat, too, is separate from other species. They migrate only short distances, often, as in Iceland, just down their breeding stream or river to the sea, where they gather off the outermost reefs and skerries of the coast, bobbing about in the rough surf, completely at ease in the largest of waves. Their food at this time includes various molluscs and crustaceans growing on the rocks. With a mere flick of its bill and head, the Harlequin can detach even the coat-of-mail shell, which a human collector can barely remove with a hammer and chisel.

The nest is always placed within a few feet of the stream and the young are immediately taken into the water by the female. By this time the males have already left the breeding area and are on their way down to the sea where they will moult and then remain there for the winter. The female and her brood make their way slowly after them, sometimes staying in one stretch for a few days, but more often allowing the stream to drift them gradually down towards the sea. By using the stream as a highway their movements are not limited by lack of flight as with other ducks. Instead they can reach the sea well before there is any chance of autumn frosts. The

An unusual scene, a male Harlequin Duck escorting the female and brood. He normally leaves all the work to his mate.

The breeding and winter ranges of the Long-tailed Duck [Old Squaw].

female stays with the young until they are fledged then she goes through the annual moult before joining the flocks of males on the sea.

Long-tailed Duck [**Old Squaw**]
Clangula hyemalis
One of the hardiest of ducks, the Longtail breeds in every arctic land, right round the pole, and often winters on the edge of the pack-ice, moving south only as far as it is pushed by the ice. It is probably the most numerous of the arctic wildfowl, though the Eider must run it very close. There are of course no complete counts of either species, but the Russians have estimated that in the western part of the Soviet arctic alone, there are perhaps 5,000,000 Longtails.

The breeding season display is one of the most stirring sights and has one of the most evocative sounds of the arctic. The male courts the female by indulging in an aerial chase during which it gives tongue to a most thrilling call. Words are inadequate to convey its wildness and music, but 'ai-aidelay, ai-ai-aidelow' is about as near as one can get. Not infrequently the female flies almost vertically upwards and the pursuing male seems to stand on its long tail as it follows her. The nest is placed in a sheltered hollow on the ground or sometimes in a crevice in a rocky outcrop, but never far from water. The clutch is fairly small, six to eight eggs, which the female incubates, while the males gather in flocks of several hundreds, and resort to the coast for the duration of the moult. When they hatch, the female takes her brood onto the water and there they live on insect larvae and small fish.

Although some of the birds winter very far north, in the nearest open water they can find to their breeding grounds, others migrate long distances. One major northern area is the rich waters off the southwest coast of Greenland. Here many hundreds of thousands of ducks of various species, and perhaps millions of auks, spend the winter. Ringing has shown that Longtails from Canada, both coasts of Greenland, and Iceland, all favour this area, which inevitably leads to mixing of populations. Some ducklings ringed in northwest Greenland in 1947 produced recoveries in subsequent years in two widely separated areas: one bird was breeding in the Mackenzie Delta region of arctic Canada, several thousand miles to the west, while another was shot during the winter in the Baltic Sea in Europe, about as far to the east. This mixing prevents the evolution of different races, so that there are no apparent distinctions between the birds from different areas. Longtails can also be found wintering in flocks of several hundred, or sometimes solitary, on the coasts of North America and Europe. Often rough weather will bring them close to the shore.

Barrow's Goldeneye
Bucephala islandica
Barrow's Goldeneye reaches the arctic in Iceland, southwest Greenland, and Labrador, and in winter it only moves as far as the coasts of Iceland and Greenland, and south to Nova Scotia and the St Lawrence in Canada. Another breeding population exists in Alaska and British Columbia but does not extend into the arctic on this side of the continent. The birds nest in trees in the southern parts of their range, but north of the tree line they find holes in rocks. In Iceland their sole breeding area is around the famous locality of Lake Myvatn, and here they nest in holes in the old lava flows that surround the lake. The nest itself may be a foot or more into the hole or crevice and the only material is the copious down plucked by the female. Unfortunately escaped mink have gone wild in the area and those sites are no longer as safe as they were, and numbers

The bold patterning of the Barrow's Goldeneye ducklings breaks up their outline and helps to conceal them from predators.

are thought to be declining. In some places Barrow's Goldeneyes have been persuaded to use man-made nest-boxes, just as their boreal relative the Common Goldeneye does. They will even use boxes attached to houses, or holes under the eaves of the houses themselves—indeed the Icelandic name 'Husond' means 'House duck', so clearly this habit is of very long standing.

The clutch of seven to ten eggs is incubated by the female for about 30 to 34 days, a comparatively long period and one that has presumably held the duck back from further northward spread. The status of the species in Greenland has long been a mystery. Small numbers of birds are commonly seen in the west each summer but there is only one proved instance of breeding when young were seen in the late nineteenth century.

Red-breasted Merganser

Mergus serrator

This bird has a wide distribution, from Alaska across Canada to Greenland, thence to Iceland, Scandinavia and much of northern Europe, and the northern parts of Russia. It breeds in the arctic regions of most of this range, and is the only purely fish-eating duck to do so—its long bill with a serrated edge along both mandibles is ideally suited for grasping such slippery prey.

The Red-breasted Merganser is a marine species for most of the year, inhabiting coastal waters, particularly estuaries and fjords. It often breeds on the coast, too, choosing a sheltered site in thick vegetation or underneath a boulder or log. However some move away from the sea to nest close to a lake or river, and in some areas there has been a considerable spread inland in the last fifty years. The clutch of seven to twelve eggs is incubated by the female for about four weeks, and the young fledge in a further seven. They are accomplished divers almost from the day they hatch and, unlike diver chicks, feed themselves from the beginning. An adult can stay under water for up to a minute and travel thirty or forty yards beneath the surface.

The Red-breasted Merganser is the only purely fish-eating arctic duck. The serrated edge of its bill helps it to grip the slipperv fish.

Common Scoter [Black Scoter]
Melanitta nigra
Surf Scoter
Melanitta perspicillata
Velvet Scoter [White-winged Scoter]
Melanitta fusca
The three scoters reach the edge of the low arctic in many parts of their ranges, in both North America and Eurasia, but true arctic breeding is confined to small numbers of Common Scoters in Iceland and occasional pairs in Spitsbergen. All three are primarily birds of more southerly latitudes, breeding on coastal islands and the bigger inland lakes. They lay six to twelve rather large eggs which are incubated for about 32 days, and the fledging period, too, is quite long at about seven weeks. They rarely come to land except for breeding, resorting to the sea during the rest of the year. Very large flocks occur in some localities, particularly where there are shellfish beds in shallow water on which they can feed.

Eiders

Included in this group of ducks is one of the best known, the Common Eider, whose down provides the stuffing for the original eiderdowns and nowadays for the finest sleeping bags, and also two species, the Spectacled Eider and Steller's Eider, whose biology is still relatively unknown. All are arctic, only the Common Eider occurring outside as a breeding species. However from its circumpolar distribution it is clear that it is a truly arctic species that has been able to adapt to conditions further south into which it has successfully spread, rather than the other way round. They are all animal feeders, taking molluscs and crustaceans from the bottom of the sea or lakes. In winter they gather in large flocks around the coasts, some migrating long distances, others remaining on or close to the breeding grounds if conditions permit.

They nest on the ground, often in colonies and sometimes completely in the open, but more often in the shelter of vegetation or rock. The clutch is only four to six eggs, but these are large. When the young hatch they are taken immediately to water, usually the sea, where several broods may combine into one large crèche looked after by a few but not necessarily all the females. Here there is some safety in numbers, and also the birds are away from the freshwater areas that might freeze up before they are fully fledged. The breeding males are remarkable for the moult migrations they indulge in after their responsibilities have ended.

Common Eider
Somateria mollissima
There are at least four races of Common Eider which occupy different parts of the arctic, and together they complete a circumpolar range. The separation into subspecies is the result of the non-migratory habits of some of the birds: by remaining in one area, minor variations in size and plumage detail have been able to evolve. The non-migratory Hudson's Eider is slightly larger than the other races and lays rather larger eggs. This is another apparent reversal of Bergman's Rule because the smaller Eiders breed to the north. However because the smaller birds are migratory and winter to the south of Hudson's Bay, Bergman's Rule does operate in winter.

These birds are mainly colonial breeders, a habit that has been positively encouraged in Iceland where the colonies are rigidly protected. Some of the Eider farmers in Iceland even go to the lengths of erecting small flags and whirling windmills which, they claim, enourages the birds to nest in larger numbers. What it may well do is to discourage predators, but then the farmers are doing this in any case. Some of the largest Icelandic colonies have as many as 6,000 nests whereas without protection this number is rarely attained, a few hundreds being more normal. The total Iceland population is put at about 500,000 pairs. The Icelanders gather the down from the nests, usually once about half way through incubation, and again after the eggs have hatched. After cleaning, the down is sold for sleeping bag fillings. Formerly it was laboriously cleaned by hand, but now machines do the job far quicker and just as well.

Colonies in the true wild are nearly always on islands or in other sites where the birds will be safe from arctic foxes. There have been a number of instances recorded when drifting ice has formed a bridge between the mainland shore and an offshore island allowing a fox to get onto the island, there to virtually wipe out all the ducks. One colony of about 120 pairs survives on the mainland of Spitsbergen because the birds have chosen to breed beside a meteorological station. The station staff shoot any arctic fox or Glaucous Gull that they see hanging around, thus repaying the trust that the birds have put in them. In past days, other colonies have been decimated by excessive egg-collecting by seamen and explorers.

Eiders breeding in the arctic suffer from similar variations in success that some of the goose species do. However studies on the duck in more southerly latitudes, for example in northern Britain, have shown that very wide variations in the number of young reared from year to year are normal even there, with really good ones coming along only once in every four or five. The reasons are not clear although, unlike the arctic, the birds do seem to lay normally most years, and the heavy losses are caused to the ducklings after hatching, probably because of some shortage of food. Further north, poor breeding

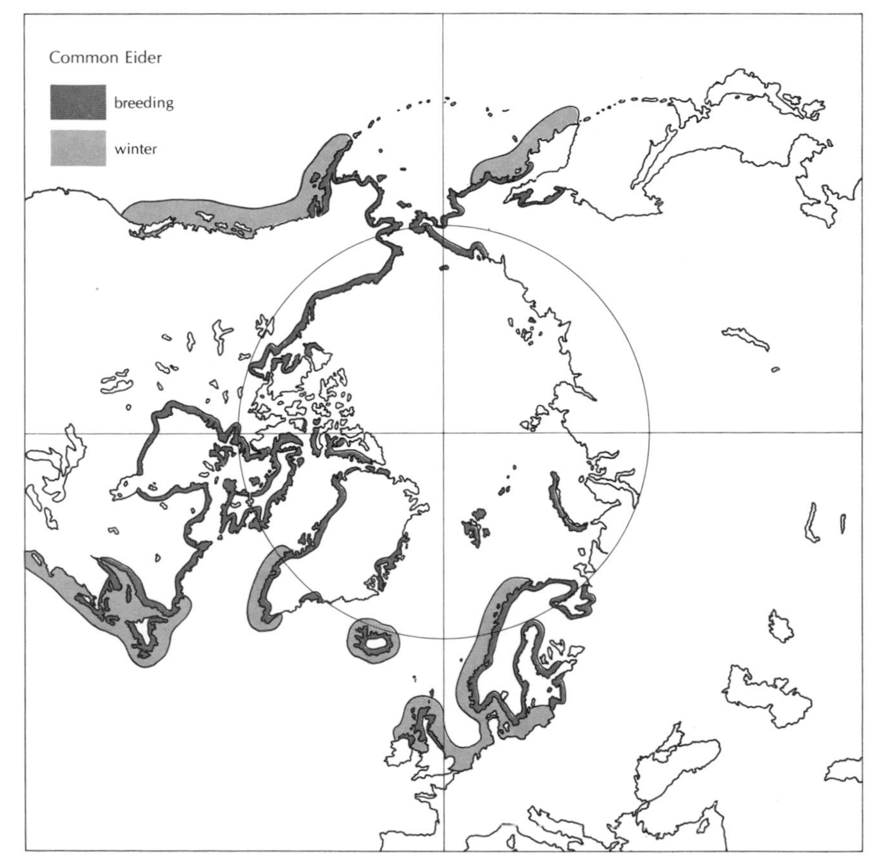

The Common Eider's breeding and winter ranges.

years are as likely to be caused by adverse conditions at egg-laying time as later after the hatch. Severe spring conditions have also been known to kill many adults. In 1964 the Beaufort Sea off Alaska remained frozen far longer than usual and very large numbers of Eiders and King Eiders that had moved into the area prior to nesting on the adjacent land died from cold and starvation. It was estimated that as many as 100,000 of the two species died, or about 10% of the total population of the area.

Spitsbergen Eiders breed in small colonies on little offshore islands and rocky skerries. Here they place their nests among rocks and boulders, gaining at least some protection from them. In a few localities the nests are placed under the huge driftwood timbers that litter the shores. Many of these have been moved well above normal tide levels by winter storms, and they form protective if unusual sites in this treeless land. The female lays three to five eggs, compared with with five to seven in more southerly latitudes where there is more food available both for the egg-producing female and for the rearing of the young. The female alone incubates and there are many authentic records of her sitting continuously for the full four week period, not coming off even once for food or drink but living on her fat reserves. If she does leave the nest, perhaps just for a drink, she carefully covers the eggs with the copious down she has plucked from her breast, and if scared off the nest, she often squirts the evil smelling contents of her caeca over the eggs. This is much more unpleasant than the normal droppings produced, and may well have the effect of discouraging any predator. When the young hatch they frequently gather into crèches of up to a hundred, looked after by a few females.

King Eider

Somateria spectabilis

It is often said that the rigours of the arctic climate have done away with the need or the time and energy for brilliant plumages and elaborate displays, but surely the colourful King Eider male disputes this. These birds have a virtually circumpolar distribution and hardly breed outside the arctic at all. In winter, too, they stay north to the limit of open water and come no further south than they apparently have to. Their wintering areas are not fully known although concentrations occur off southwest Greenland and in the White Sea. Stragglers reach south into

The male King Eider is an exception to the rule that arctic species do not bother with colourful plumage.

temperate waters and even occasionally stay to breed well to the south of the normal range. Hybrid pairs between a male King Eider and a female Common Eider have been recorded many times in some of the Icelandic colonies.

The King Eider performs one of the more spectacular and complete moult migrations that has so far been discovered. Virtually the whole population of breeding males moves from the greater part of the Canadian arctic, both the mainland coast and the archipelago of islands, to a rather small area off the west Greenland coast, around Upernavik. Here they are joined by the King Eiders from west and north Greenland in a gathering that numbers at its peak around 100,000 birds. This enormous concentration includes a great many immature non-breeding one year olds. Some move straight from the wintering grounds, which lie to the south of the moulting area, others migrate part of the way to the breeding areas in spring, but before they reach them, turn about and move back south.

Almost exactly half way across the Canadian arctic, at Victoria Island, there is an invisible divide, to the east of which the King Eiders move eastwards to moult off west Greenland, and to the west of which the birds go west to somewhere on the Russian side of the Bering Straits. The exact location is not known but enormous numbers of male birds migrate past Point Barrow in Alaska each July heading west to the other side of the Straits. Those breeding in the Soviet arctic also moult migrate, either westwards to an area around Kolguev Island in northwest Russia, or eastwards to the northeastern part of the Siberian coast.

King Eiders are solitary breeders, often nesting well inland, and sometimes up to half a mile from the nearest water. More usually however they are close to a pool, if not the sea, to which the young are taken on hatching. Such scattered breeding parallels that of the Canada Goose where the safety offered by a colony is balanced unfavourably against the safety of being well dispersed and therefore less likely to attract the attention of a predator. In some areas several broods may end up on the same lake or area of sea, in which case they will gather into a crèche like those of the Common Eider.

Spectacled Eider

Somateria fischeri

One of the least known of arctic breeding ducks, the Spectacled Eider has a breeding range confined to Alaska and northeastern Siberia. In the latter area it is relatively common in some parts and 17,500 pairs were estimated for the delta of the Indigirka river in 1971. Here the birds breed in the extensive coastal bogs which are liberally endowed with shallow pools. They arrive on the breeding grounds already paired, having waited offshore until the pools have melted.

The nests in the Indigirka delta are either scattered thinly over the tundra in a similar manner to the King Eider, or are gathered in

A female Spectacled Eider settling on her eggs. Both sexes have the extraordinary eye-ring which gives the species its name.

Two male Steller's Eiders in Varanger Fjord, northern Norway. The true status of the species in this area is still not known.

small colonies of up to dozen on small islets in the lakes. It was found that these colonies were often only two to twenty feet away from the nests of breeding gulls and terns. Normally the gulls, many of them Glaucous or Herring, would be classed as predators of the ducks and indeed several of the Spectacled Eider nests on these islands were destroyed by them. However the very closest nests were usually left alone. The explanation is that a pair of gulls defends the immediate area of the nest against other gulls of their own species, as well as against predators such as skuas and foxes, and so an Eider's nest within this ring of protection would benefit because it would have only two gulls in a position to predate it. Such breeding relationships have been recorded many times, and not just in the arctic. Spectacled Eiders that lost their eggs shortly after laying were able to lay repeat clutches, but if these were lost their breeding effort ceased. Not only would their energy reserves be too depleted but there is an obvious limit on how late it is worth trying to breed in order to complete the cycle before winter comes.

Little is known about the migrations of the Spectacled Eider. Small numbers certainly move westwards along the north Russian coast to appear around the northern tip of Norway and even straggle down into northwest Europe. Most seem to winter in the seas south of the Bering Straits, probably staying close to the edge of the pack ice.

Steller's Eider
Polysticta stelleri
Steller's Eider breeds in a comparatively restricted area of northeast Siberia and western Alaska, and winters in an even more circumscribed region around the eastern Aleutian islands and the Alaskan west coast. This is also the area where the males and immatures come for the moult so it is carried out actually on the wintering grounds, and no further migration is made after their new feathers are grown, unlike, for example, the King Eider. However in some years the birds do not arrive until November and it is assumed that they have moulted elsewhere, presumably closer to the nesting grounds. These occurrences coincide with late springs and delayed breeding seasons, and such a setback is probably the reason why the birds fail to reach the normal moulting grounds in time.

These birds are rarely seen away from the arctic. Small numbers summer around the northern tip of Norway, and there have even been one or two breeding reports from the area, but this is an isolated group of a few hundred birds thousands of miles from the rest of the population, and its exact status is not clear. The breeding habits of the species have not been studied in detail, but are essentially similar to the other eiders.

4 Waders or shorebirds

The waders or shorebirds form the largest single group breeding in the arctic, just outnumbering the waterfowl. Collectively they occupy a niche that is dependent in part on water but may also extend to very dry stony or heathy land (from where they return to wetlands for the winter). A few nest and rear their brood in rather arid conditions but most take their young to the edges of pools or to marshes where there is an abundance of food. Hardly any species are colonial—they usually breed in low densities scattered over very wide areas of tundra. Waders are normally animal and insect feeders but in the arctic may have to feed on plant material for periods, particularly on arrival in spring. Their breeding cycle is comparatively short with the incubation and fledging periods rarely totalling more than six or seven weeks combined. Nevertheless they start their egg laying very quickly after the spring thaw and do not apparently have time for a replacement clutch if the first is lost.

The various species differ greatly in the manner in which the parents share the chores of incubation and rearing the young. In most, the adults leave the breeding grounds well before the young, often as soon as the latter have fledged. There are two reasons for this. First the food supplies remaining on the breeding grounds are left to the young for them to use to put on the necessary premigratory fat between fledging and their first migration. Second the adults need much richer feeding than can be provided in the arctic for their annual moult. Again there is great species variation in its timing: it can be carried out at the wintering grounds, at some earlier migration stopping place, or at a mixture of the two. The distances travelled by waders on migration are, on the whole, much greater than even those of the wildfowl. Flights from the top of one continent to the foot of another are commonplace, but such journeys involve a number of traditional migration resting places, and the protection of these, many of them estuaries and mudflats threatened by reclamation and/or pollution, present many major conservation problems today. It is not enough just to protect the breeding and wintering grounds for these birds; the intermediate stops are equally important.

As far as possible the normally accepted classification of the waders or shorebirds is followed here, but it has been departed from slightly in order to be able to group more closely birds of about the same size. Clearly there will be more similarities in adaptations to the arctic in such a group than if some were large and others small. The only species that is moved a long way from its normal position is the Turnstone, here grouped with the small waders, although these are not its closest relatives.

Golden Plovers breed in the arctic on dry upland tundras, while further south they inhabit moorlands and heaths.

Plovers

There are four species of plover in the arctic, though only one, the Grey, is exclusively an arctic breeder. The other three all have ranges covering areas far to the south, usually involving separate subspecies in different parts of the range. Like virtually all waders, they nest on the ground, forming a shallow scrape by turning movements of the chest and belly. This rudimentary nest is usually given a little lining, small stones in dry areas, the leaves of plants in wetter, more vegetated sites. Plovers breed principally on dry upland regions of tundra, just as those breeding to the south favour heathy moorland or stony beaches. A similarity shared by all four species is the predominance of black and white in their breeding plumage. This is used in display but the brightest patterning is only seen in flight or on the under parts of the body, so that when incubating even the most conspicuous birds become beautifully camouflaged. The Ringed Plover is the only one of the four that does not have a different plumage for summer and winter.

Golden Plover
Pluvialis apricaria
American Golden Plover
Pluvialis dominica
Although almost certainly two distinct species, it is convenient here to deal with these birds together. They replace each other round the pole, the American Golden Plover breeding in northern Alaska and most of northern Canada, the European Golden Plover breed-

The breeding and winter ranges of the Golden Plover and the American Golden Plover.

Several species, including the Long-tailed Duck [Old Squaw], have a complete circumpolar range, occurring in all arctic countries and island groups.

The female Common Eider lines her nest with down to insulate her eggs from the cold and damp. In Iceland this is collected for sleeping-bag fillings.

In winter the Lesser Golden Plover loses the black-and-white underparts of its summer dress and the golden flecking on its back.

ing in Iceland, Scandinavia, and Siberia east to the Yenesei. A subspecies of the American bird, usually called the Lesser Golden Plover, fills in the gap from central arctic Siberia east to the west coast of Alaska. Both Golden Plovers nest in the drier areas of the arctic, particularly on well vegetated slopes and high uplands, in much the same sort of situation as they do further south in, for example, northern Britain and Scandinavia. The nest itself is a shallow scrape lined with leaves, and the usual clutch is four eggs. In the American species both parents apparently share the incubation of about twenty-six days, while in the European the female is thought to do much the larger share, but in each case both parents help to rear the young. As soon as these are fledged, in the first half of August, the parents depart from the breeding grounds and begin their southward migration, with the young following after in a further two or three weeks.

In Europe Golden Plovers winter no further south than the Mediterranean and often as far north as the winter snows allow. Flocks of several hundreds or even thousands are commonly seen on inland plough and grassland, the species having no particular affinity for the traditional wader habitat of mudflats and estuaries, except in severe weather. However the American species migrates to South America, wintering on the plains of Brazil south to Argentina, and the Lesser Golden Plover winters in eastern Asia and Australia, and in New Zealand. Thus the European species has much less far to travel and consequently spends relatively more time on the wintering grounds. It is here that it carries out its annual moult, whereas the American species may start during migration.

Grey Plover [Black-bellied Plover]

Charadrius squatarola

Among the plovers, this is the only completely arctic breeding species, with a rather discontinuous but circumpolar range, having separate breeding areas in Siberia and Canada. The two populations have quite distinct migration routes and winter quarters but have clearly not been separated for long enough for any physical differences to have evolved. There are no large concentrations of Grey Plovers in the arctic, for the pairs breed in well scattered fashion over the tundra. Their choice of nest site is catholic, and ranges from coastal marshes to inland gravelly plains, rocky semi-deserts, and uplands. The usual scrape is lined with a few small stones or leaves and the clutch is normally four, with an incubation period of about 27 days, but the share that the parents take has not been determined. Some authorities say that only the male incubates, while others state that both do, and there is equal disagreement over the rearing.

The parents usually complete their migration before moulting. Occasionally, however, birds arrive on British wintering grounds with some of their wing feathers already lost and the new ones grown. It is not known whether they started their moult on the breeding grounds or at some intermediate stopping place, but it is possible that in a very early season, or perhaps one in which the adults failed to breed, there is the time and the food available for the moult to be started. The adults arrive in Britain in August but the young do not reach the wintering grounds until the second half of September, indicating a much later departure. Here, Grey Plover can be seen throughout the year, the summering flocks being assumed to be

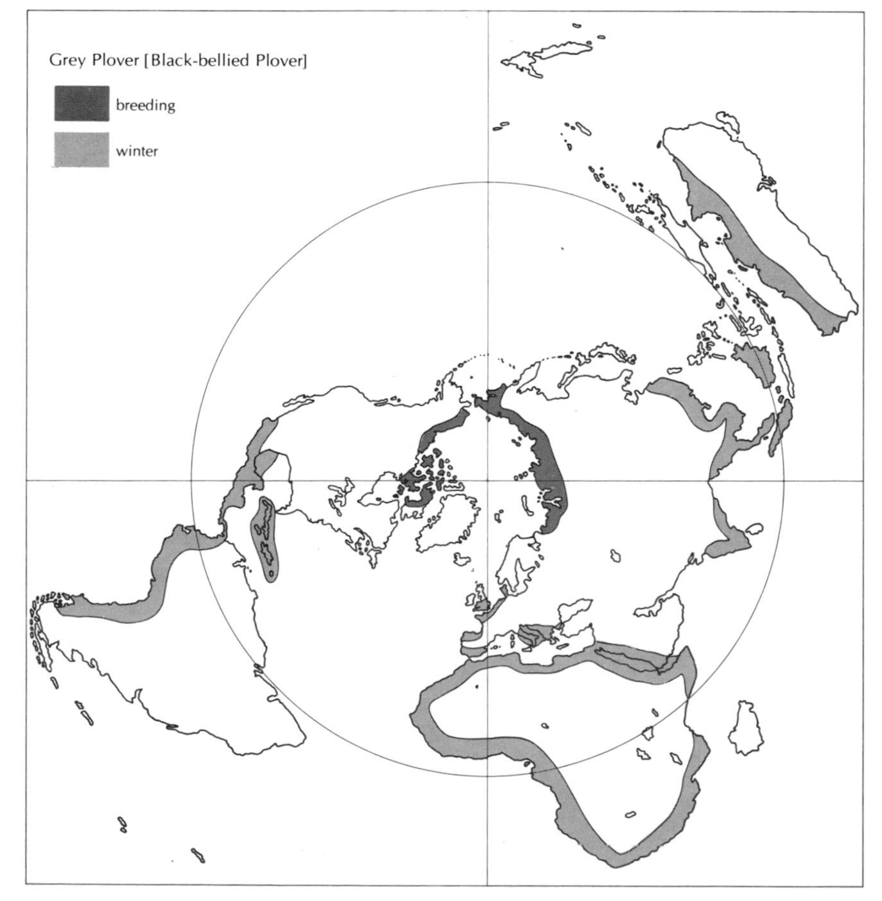

The Grey Plover's [Black-bellied Plover's] breeding and winter ranges.

composed of first year, non-breeding birds. Eurasian Grey Plover are almost invariably found on coastal mudflats in winter, but in North America they occur with some regularity inland on wet, short-grass fields and beside very large lakes and rivers. The migration of this population takes at least some of them south as far as Brazil, while others winter on both seaboards of the United States.

Ringed Plover including **Semi-palmated Plover**
Charadrius hiaticula
Taxonomists puzzle over this species, calling it one, or splitting it into two with equal facility. The only physical distinction between them appears to be the minute webbing between the bases of the toes, which occurs between all three toes in the Semi-palmated Plover and only between two in the Ringed, but this can hardly be classed as a good field distinction. The ranges of the two are almost complementary with a small but uncertain area of overlap in western Baffin Island. Otherwise the Semi-palmated Plover breeds right across North America from Alaska to Newfoundland, while the Ringed Plover breeds in the extreme northeast of the arctic Canadian islands of Ellesmere and Baffin, then through Greenland, Iceland, and across arctic Eurasia to the Chukotski peninsula. The two together thus have a complete circumpolar distribution, and their almost total lack of overlap is further argument for treating them as one species, as is done here.

The Ringed Plover breeds quite far south, for example all round the coasts of Britain, and also in North America, particularly down both coasts. Migration takes the majority of

The Ringed Plover makes a shallow scrape for a nest in barren rocky valleys or on shingle storm beaches.

The breeding and winter ranges of the Ringed Plover.

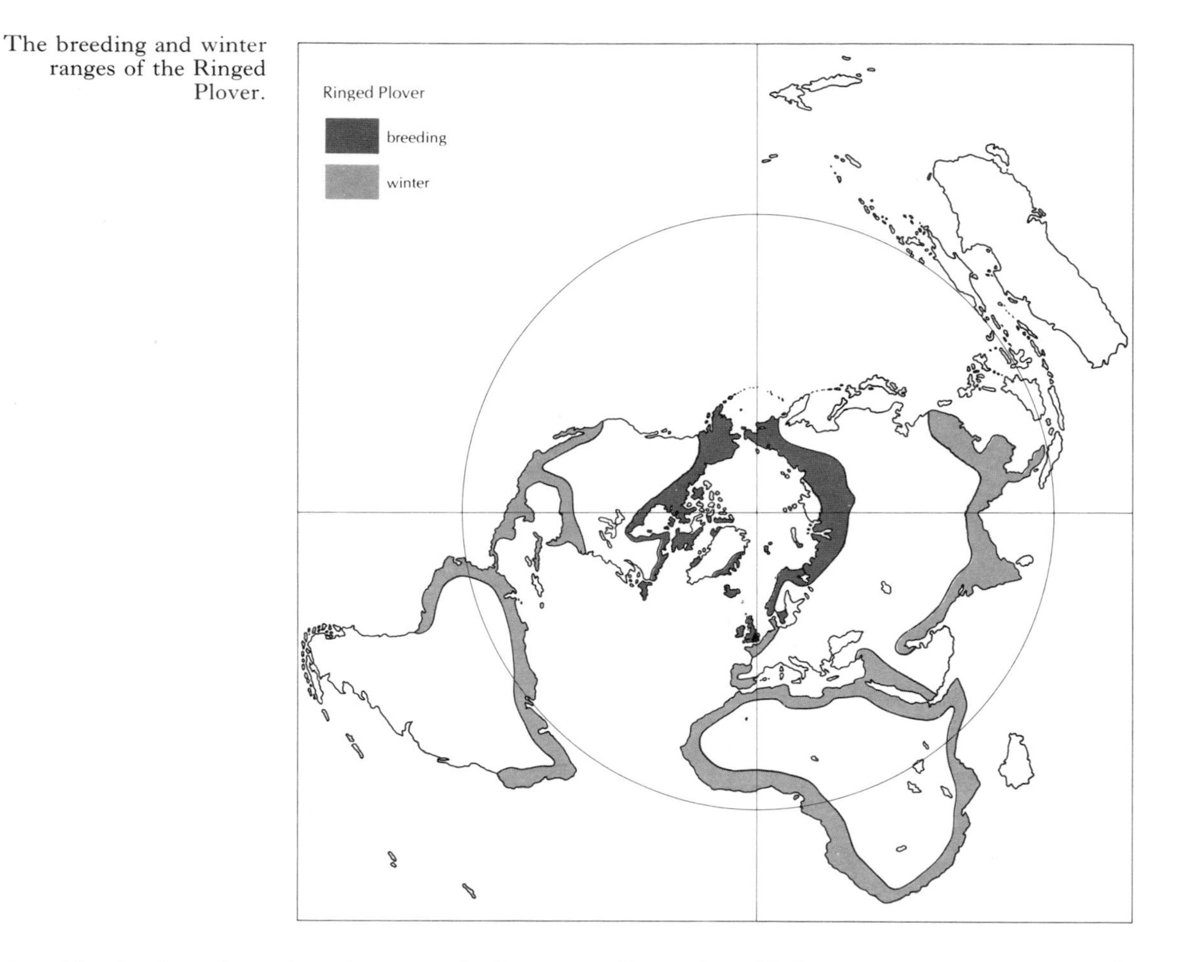

the North American breeders south into South America, with the exception of those breeding in the far northeast (the true Ringed Plovers according to some) which migrate to winter in Europe. The Eurasian birds winter from the Mediterranean south into Africa, and in India and China. There are a number of different races, mostly separable only on size. In general the larger breed to the south of the smaller, but as the latter perform a leapfrog migration, Bergman's Rule is obeyed in the winter.

The male arrives on the breeding grounds ahead of the female and takes up a territory. The habitat is far more barren and much drier than for almost any other wader, with the species preferring stony plains and rocky valleys where little or no vegetation grows. It has been found as high as 3,000 feet but it also breeds on shingle storm beaches on the coast, not very different from its coastal habitat in Britain. However in the steep rocky valleys of, for example, East Greenland, it may be the only bird encountered, apart from an occasional Snow Bunting. Walking down one of these valleys, nesting pairs can be found every mile or so, the off-duty bird pursuing one with its plaintive two-note anxiety call. No other wader will be seen until the more marshy habitats in the bottoms of the much larger flat valleys. Although these marshes are richer in insect food than the rocky valleys, there is obviously sufficient in the latter for well scattered pairs and their young. The usual clutch is four eggs, though up to six have been reported in some areas. This, if true, would presumably be an adaptation to variable weather conditions during the breeding season, allowing for maximum production in good years. Their incubation period is twenty-three days, and the parents share this between them, as they do the rearing of the young. The adult birds leave the breeding grounds in late July or early August, while the young remain for up to another two months, rather longer than most other juvenile waders, with the last departing at the end of September. Ringed Plovers from Greenland and Iceland pass through Britain on their way further south, and many pause in estuaries for about three weeks during the late summer and autumn. They use this period purely for feeding up and laying down stocks of fat, delaying their annual moult until they reach their winter quarters.

Small Waders

The small waders are made up of twenty species—twelve that are truly arctic in their range, plus another eight that are only secondarily so. They are all closely related with the exception of the Turnstone which is brought in here because its size, and therefore adaptations to the arctic, are more similar to the small waders than to its larger, if closer, relatives. One of the most striking things about these birds is that due to minor variations in habitat requirement, several of them are able to live in close proximity. No two species can occupy exactly the same niche all the time, but in certain conditions, as for example in periods of superabundance of food, then they can overlap without competing. For example a study has been made of four small waders breeding in the same area of arctic Alaska (their range also overlaps through much of arctic Canada): the Dunlin or Red-backed Sandpiper, and Baird's, Semipalmated, and Pectoral Sandpipers. The four species arrive on the breeding grounds in June, at which time much of the lower lying marshy land is under snow and then, as the snow melts, water. The waders therefore resort to the drier uplands. Here the Baird's are found in the driest sites, while the Semipalmated and Pectoral Sandpipers are in the wettest, with the Dunlin in between. The Dunlin and Pectoral Sandpiper are in well vegetated areas, while the other two favour more barren parts. At this time the Semipalmated Sandpiper is feeding on surface insects, but the Pectoral Sandpiper (in similar habitat) and the other two species concentrate on insect larvae, obtained by probing just below the surface. In July, when the surface insects are at their most plentiful, all four species feed on them, but competition is avoided mainly by their superabundance. Additionally the Baird's Sandpiper remains on the drier uplands to breed, separated from the other three which have moved down to the wetter marshes. Then in August, the Dunlin and Pectoral Sandpiper return to feeding on sub-surface insects, while the other two continue feeding on surface insects.

Knot
Calidris canutus
These birds are much less well known than many of the arctic waders. Their breeding range, exclusively within the high arctic and very localised, is in northern and eastern Greenland, Spitsbergen, a small area of central northern Siberia, northern Alaska, and the most northerly of the Canadian arctic islands. Over the known parts of this range they occur in scattered pairs with no concentrations to permit easy study. The Knot arrive on the breeding grounds in flocks in late May and early June. Within a very few days they are distributed in pairs over the tundra—an average density in one area of arctic Canada was one pair every 250 acres. The breeding habitat consists of high and dry inland slopes and plains, often with scant vegetation, and right away from marshy ground. However there must always be either

The Knot's breeding and winter ranges.

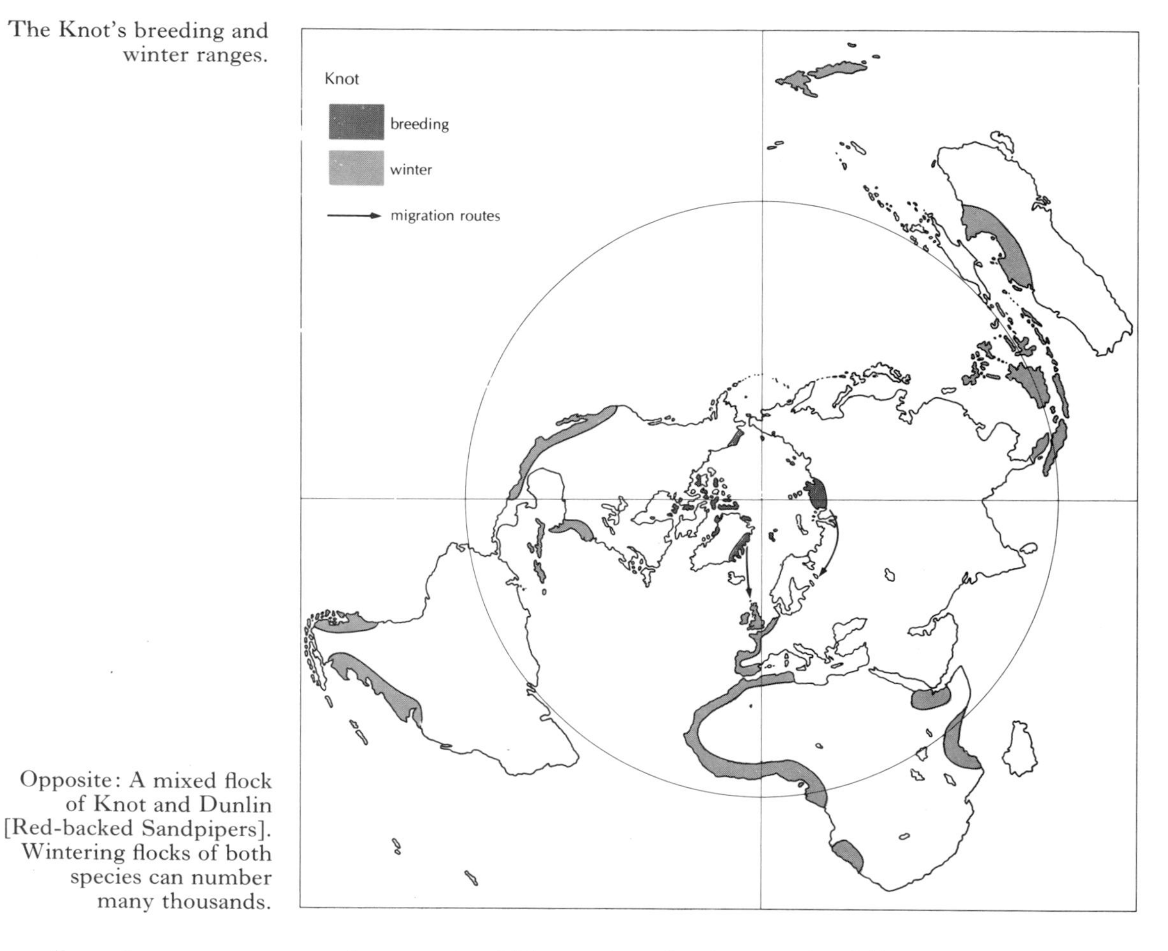

Opposite: A mixed flock of Knot and Dunlin [Red-backed Sandpipers]. Wintering flocks of both species can number many thousands.

small pools or streams nearby round which the parents can feed and to which they can take their young; sometimes, though, the young are led quite long distances to reach wetter marshland.

The nest is the usual shallow scrape, and the clutch is of four eggs. There has been a certain amount of dispute about the part played by the sexes in incubating and rearing, with some authorities saying that both parents incubate and rear, while others think that only the male rears, though both incubate. It has now been established that neither idea is incorrect, in that although both parents incubate and begin to rear the brood, the female leaves to start her southward migration after the first week while the male carries on until the young are fledged. The incubation period has only recently been accurately assessed at 21 to 22 days. The breeding season is timed so that the hatching of the young coincides with the greatest emergence of adult chironomids. The gradual decrease in insect food from this peak is counteracted by the wandering of the family parties over the tundra and by the early departure of the adult females. Indeed the parents rely heavily on plant material throughout the summer, and it forms their major food in the period immediately after arrival. As insects become available so the adults turn more and more to them, but are still eating up to 50% plant

food at the height of the insect hatch. Thus competition between the parents and their young is reduced.

The Knot breeding to the north and west of Hudson Bay, including Alaska, migrate south to the Mexico Gulf and into South America. However the birds from most of the Canadian islands join the northern Greenland birds in migrating to Europe where they meet the Spitsbergen and Siberian populations. The largest concentrations are found in northwest Europe and down the Atlantic coast as far as West Africa. As with virtually all the waders, the adults leave the breeding grounds first, followed about three weeks later, in the last half of August, by the young. The rapidity of these migrations and the distances travelled were dramatically demonstrated when a large number of Knot were caught and ringed on the coast of the Wash, in eastern England, at the beginning of September 1963. No less than four birds were subsequently recovered in Sénégal, all within five weeks of ringing, while one bird reached Liberia only eight days after it was in England. And these were young birds that had already travelled, within the previous week or at most fortnight, from somewhere within the high arctic. Further ringing has confirmed that most young Knot winter to the south of the adults. The wintering flocks can number several tens of thousands, and the sight of them flying through the air, wheeling and turning together, is a wonderful spectacle. They feed during low tide on estuarine mudflats, but as the tide comes in they are forced into flocks, often packed tight on to some small ridge of shingle, where they roost until the water recedes.

Sanderling
Calidris alba
The Sanderling shares with Temminck's Stint the distinction of one of the more unusual and interesting adaptations to breeding in the arctic. Within the last ten years it has been discovered that Sanderlings breeding on Bathurst Island (in the Canadian arctic) may lay two clutches in a summer, not as in many species in temperate lands, laying the second after the first brood has been reared, but laying them one after the other, with each parent responsible for one of them. This extraordinary behaviour is not followed by all Sanderling, but it is certainly not just an aberration of a few.

The adults arrive already paired on the breeding grounds in early June, and then adopt one of two distinct breeding patterns. In some nests the usual procedure is followed, incubation commencing immediately the full clutch of four eggs is laid, but in others, the eggs are left uncovered for between five and six days before incubation begins. During

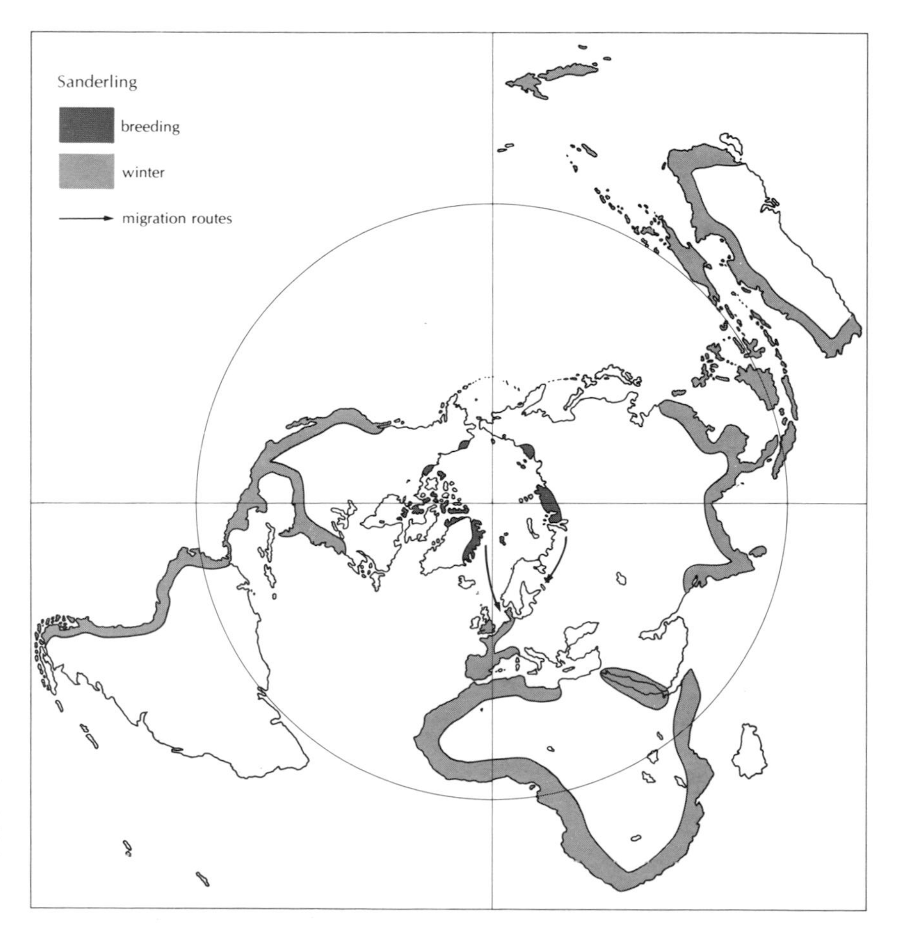

The breeding and winter ranges of the Sanderling.

Opposite: Grey Plovers [Black-bellied Plovers] moult from their grey winter plumage into a more distinctive breeding dress. However, the incubating bird is still well camouflaged.

The Purple Sandpiper sitting on its nest has allowed the photographer to approach within a few feet. It is just as indifferent to man in the winter.

this period the female is engaged in laying a second clutch at some distance, perhaps as much as a mile, from the first nest. These are then incubated by the male, while the female returns to the first nest and incubates the eggs she has laid there. Among the evidence for this behaviour is the fact that on Bathurst Island at least, and in some other parts of the arctic too, only one adult has been found either incubating or rearing the young, and that some of these have been proved to be males, others females. This is not universal, however, as in East Greenland there are records of both parents sharing the duties of incubation and rearing. Another strong proof comes from the dissection of two female Sanderling, shot for the purpose on Bathurst Island. They both had eight egg follicles in their ovaries instead of the expected four, suggesting that each had laid eight eggs within a short period. Indeed they were thought to be physically capable of laying even more, perhaps as many as twelve.

The most likely reason for this intriguing adaptation is that it enables the birds to multiply the traditional clutch in favourable years, thus helping to counterbalance the effect of breeding failures in bad years. It also means that the bird could relay if either of its first clutches were destroyed by a predator, a possibility not proved for other arctic waders. Surprisingly, all this is not beyond the bird's energy resources, because the Sanderling's eggs are smaller in proportion to its body weight than for most other waders. How widespread this habit is remains to be seen. Certainly some East Greenland Sanderling do not follow it, and scepticism has been expressed about the Bathurst occurrences, with the suggestion that it is an isolated freak. There is certainly one very puzzling aspect that remains completely unexplained: it would appear far more sensible for the male to incubate the first clutch of eggs laid, and the female to incubate the second, for although the first clutch will come to little harm from the weather because unincubated eggs are highly resistant to cold and damp, the danger from predation must be high as they are not covered in any way and rely solely on their cryptic colouring. The normal incubation period for Sanderling eggs is twenty-four days, with the nest always placed in quite dry situations, including stony barrens and rock desert not very dissimilar from the Ringed Plover's habitat. They have been found up to 2,500 feet above sea level.

Migration for the adults begins in the first half of August, and in late August and during September for the young. North American

Two features immediately identify a Sanderling in winter: its clean white plumage and its habit of running along the edge of the tide.

Opposite: Among the rarer arctic waders is the Sharp-tailed Sandpiper. It breeds in eastern Siberia but occasional stragglers have reached Europe, mostly in the autumn.

Sanderling migrate south down both sides of the continent, some wintering in the southern United States, but the majority going much further south to Chile and Argentina. In the Old World they winter from the British Isles and Japan, south to South Africa, Australia, and New Zealand. Ringing in Britain has shown two distinct populations involved with completely distinct habits. Birds arriving from Greenland moult almost immediately, and then stay for the rest of the winter. Siberian Sanderling, on the other hand, only spend a few weeks in Britain in the autumn, feeding hard and putting on weight, but not moulting. This they do on their wintering grounds. Sanderling flock on sandy rather than muddy shores, and in flocks of hundreds rather than thousands. They are quickly spotted by their feeding behaviour of running on twinkling feet along the tide edge, pecking up small creatures as they go.

Semi-palmated Sandpiper

Calidris pusilla

One of the smallest arctic waders, the Semi-palmated Sandpiper is confined to the New World, where it breeds from northern Alaska across the low arctic region of Canada to the coast of Labrador. Its nesting habitat is normally in low coastal areas or on wet inland tundra, usually with plenty of vegetation cover such as long grass or even dwarf willow, though nests on bare sand beside rivers and pools are not uncommon. They are invariably placed quite near to water, whether under cover or exposed.

The clutch is four eggs, and both parents share the incubation period of about nineteen days. However it is the male that has the larger share of rearing the young, although the female does not leave until perhaps the last week of the three-week fledging period. As soon as this is over the parents move to the nearest coast where not only is the feeding likely to be better than inland, but the young are left with less competition for what is available. The adults start their body moult before they migrate but do not change their wing feathers until they have reached some resting place further south. Their migration is almost wholly down the east side of North America, with even those breeding in Alaska crossing the continent rather than going down the Pacific coast. The ultimate wintering grounds are from the Gulf coast south to Chile and Brazil. (Some of the habitat and food requirements are dealt with in the introductory paragraph to the small waders.)

White-rumped Sandpiper

Calidris fuscicollis

This is another small wader confined to North America, extending further north in the breeding season than the previous species, and not wintering so far south. It likes similar habitat, though perhaps rather wetter tundra, and, without going very far inland, it also avoids the coastal strip. The male arrives on the breeding grounds first and marks out a small territory. Here he courts the female when she arrives, and within a matter of days mating will take place. The pair bond is of extremely short duration as the female carries out all the incubation and rearing of the young. She sites the nest without any regard for the male's territory, clearly showing that it does not have the normal function of providing sufficient food for the pair and the young. It can only be for the purpose of advertising the male's presence, which is more likely to succeed if he is living in one defined area rather than wandering over the featureless tundra. Whether or not the same female returns to the same male each year is not known, but there is some evidence to suggest that one male may attract and mate with more than one female. Such polygamy would obviously increase production in an area where females out-numbered males.

The female incubates the four eggs for twenty-two days. As they hatch she carefully removes each shell fragment to some distance from the nest, because the white inside of the broken shell would prove far too conspicuous for a passing predator. This is probably normal practice for most waders but it has not been recorded for all. As soon as the young are dry the female leads them to the nearest water, which can be half a mile away or more. The fledging period is about seventeen days, after which the female leaves on migration. The male, meantime, has already left the breeding grounds, doing so earlier than any other wader (mid-June onwards), having probably only spent as little as two or three weeks there. This is one of the shortest periods for any breeding bird to actually spend on the nesting area. The stresses and strains of barely completing one migration before turning round and starting back again must be set against the minimal impact that the males have on the food resources, which are strictly limited until the peak of the insect hatch in mid-July. By the end of July the females are on their way, after which the young gather in flocks on the coast and follow their parents during August.

The autumn migration takes most birds to the east coast, though some journey through the interior, to reach their southern American wintering grounds. In spring, the return migration is more overland, up the middle of the continent to the east of the Rockies and the west of Hudson Bay.

Baird's Sandpiper
Calidris bairdii
Although Baird's Sandpiper is principally a North American arctic breeding species, its range extends into both northeastern Siberia and northwest Greenland. The whole population, however, migrates south through the United States to winter in western and southern South America. The occasional bird that turns up in Europe has probably gone astray on leaving Greenland or northeast Canada, perhaps not surprising in view of the many thousands of waders which normally go to Europe from this region, including Knot, Turnstone, and Ringed Plover. These different migration routes from the same breeding area provide good evidence as to a population's origin. Thus Baird's Sandpiper must have survived the ice ages somewhere on the North American continent south of the icecaps, and then expanded north as the ice retreated. A little more successful or perhaps adaptable than some of its relatives such as the White-rumped and Semi-palmated Sandpiper, and certainly capable of living further north than these, it has managed to expand into both Siberia and Greenland. However unless it can overcome the enormous hurdle of establishing completely new migration routes and winter quarters it is unlikely to spread any further west or east.

This bird prefers drier areas for breeding and feeding than other closely related and sympatric species, so its nest site is usually on dry upland slopes or sometimes in stony areas, and only rarely among marshy tundra. It occurs both on the coast and inland. The clutch is occasionally four, but three eggs are perhaps commoner in Baird's Sandpiper than in most other small waders. Both parents share in the twenty-one day incubation and both help to rear the chicks, and as soon as this is finished they leave, usually by mid-August, while the young do not depart for another two or three weeks.

Pectoral Sandpiper
Calidris melanotos
There are a number of similarities between the breeding behaviour of this species and that of the White-rumped Sandpiper. Independently, they seem to have arrived at the same methods of adapting to the arctic. Their range, too, is rather similar, although there is little or no competition between them because the Pectoral is larger than the White-rumped, and therefore normally takes larger food items. In addition there is a difference in size between the sexes, with the male larger than the female, which also reduces food competition.

The male arrives on the breeding grounds first and takes up a territory on the relatively snow-free slopes above the still covered marshes. The females arrive shortly after and are attracted to the male's territory by a courtship display in which he greatly inflates his throat, meanwhile emitting a curious grunting sound, an action quite unlike that of any other wader. Both the male and the female rely on the territory for supplying food during the courtship and pairing period. Although it has not been proved, it is probable that the territory size is linked with food availability, and numbers can vary from fifteen pairs per 100 acres, a high density for a wader, down to as low as three pairs in a poor season. As with the White-rumped Sandpiper there are well-documented cases of polygamy.

The female makes the nest scrape at a site of her own choosing and not necessarily in or near her mate's territory. It is usually placed in a well-drained but well-vegetated tundra area, often on the slightly raised bank of a stream or pool, or occasionally on a dry hummock surrounded by marsh. Four eggs are laid and incubated for between twenty-one and twenty-three days. As the pair bond only lasts until the end of laying, the female is left to incubate and rear the young. Pectoral Sandpipers prefer grassy marshes, wet fields, and the edges of pools for winter feeding, and are not found so frequently in the traditional wader habitat of muddy estuaries.

The female Pectoral Sandpiper incubates and rears the young without any help from the male.

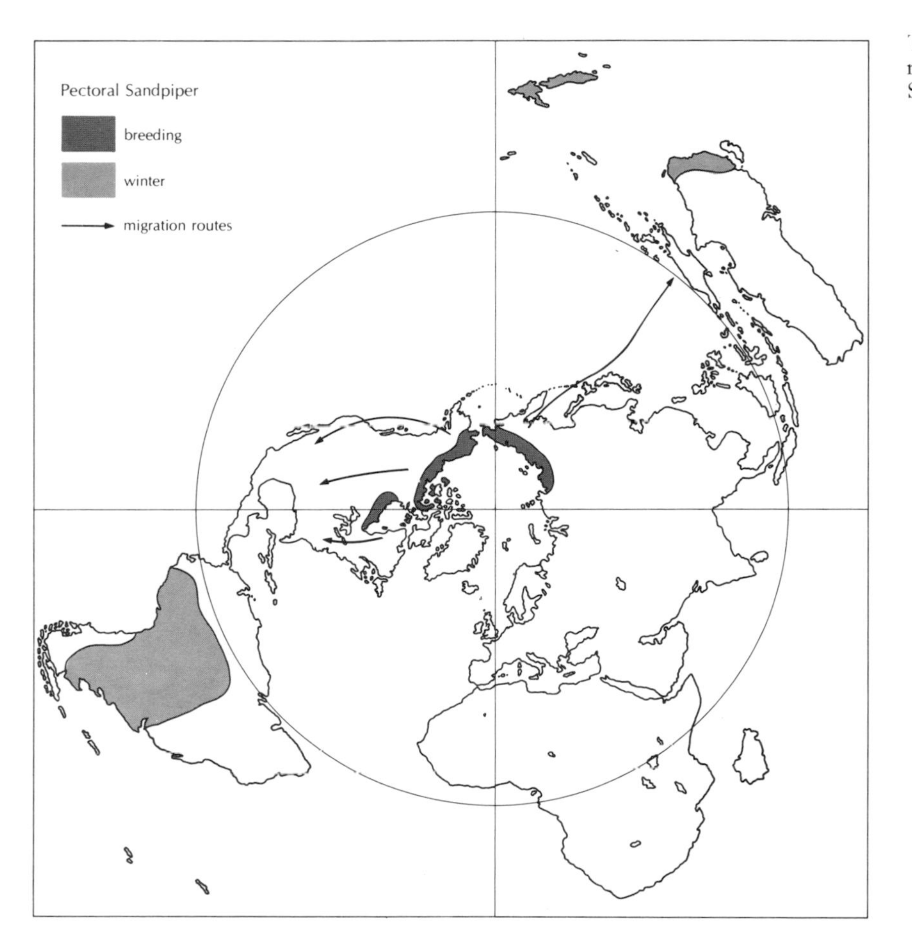

The breeding and winter ranges of the Pectoral Sandpiper

Curlew Sandpiper
Calidris ferruginea
Because it nests almost exclusively in Siberia, this species is less well known in the breeding season than many other waders. However part of the population migrates to northwest Europe on its way to Africa, and it is a familiar bird in autumn and spring in these estuaries. Those that pass through Britain in the autumn stay only long enough to put on fat for their further migration, and they delay their annual moult until they reach the wintering grounds. Numbers are usually quite small, but every so often much larger flocks appear, sometimes totalling several hundreds or even thousands. Thus in 1969, following what must have been an excellent season, there was a very large passage of mainly young birds in late August and September. It is also possible that the birds were following a more westerly course than usual, coming across Britain instead of across the centre of Europe to the Mediterranean.

The Curlew Sandpiper is another arctic wader in which the male takes no part in the incubation or rearing of the young. Like the Pectoral and White-rumped Sandpipers, it holds a territory during the courtship and laying periods, but then leaves the area. The female lays four eggs, and incubates for about twenty-two days, though it has not been precisely measured.

The Curlew Sandpiper's breeding and winter ranges.

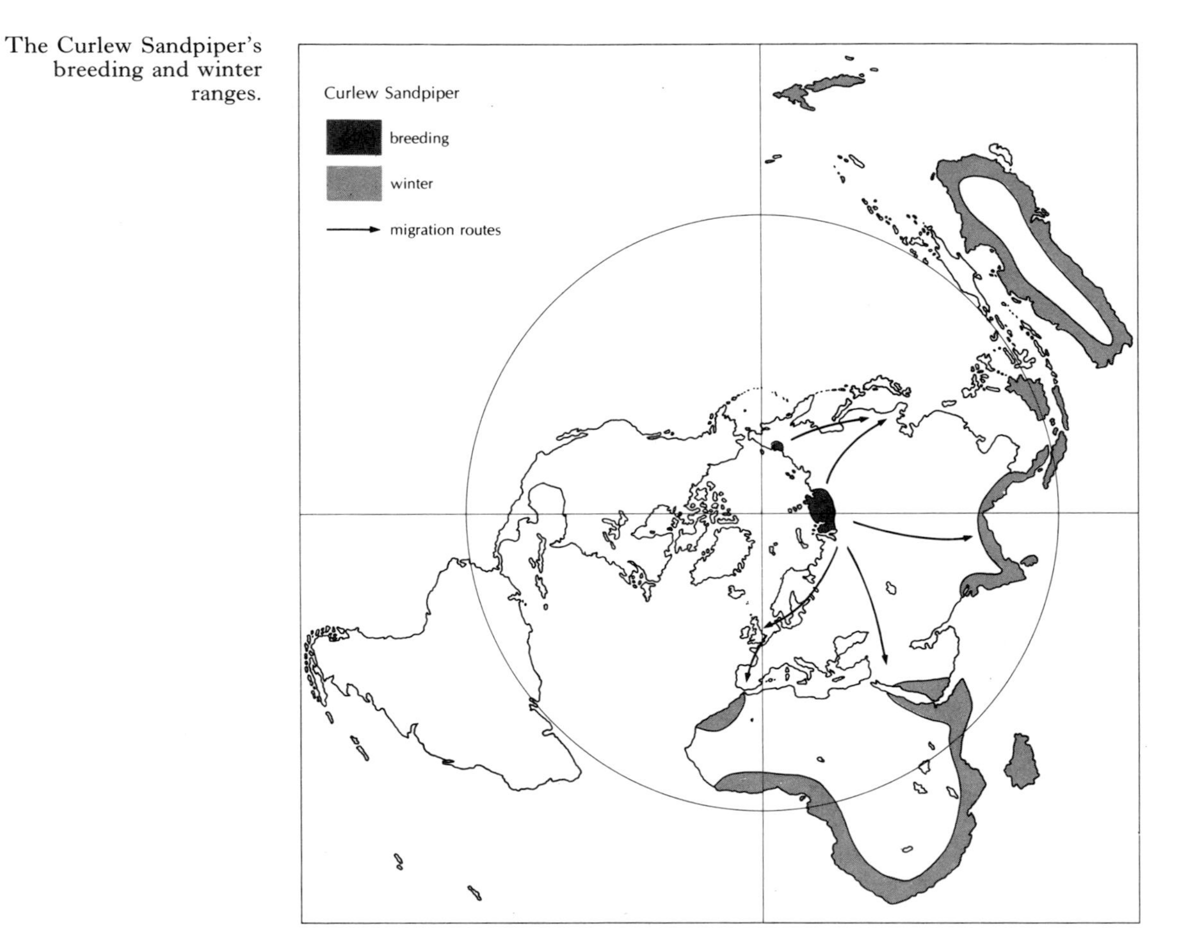

Purple Sandpiper
Calidris maritima
including **Rock Sandpiper**
Calidris ptilocnemis

The Purple and Rock Sandpipers are often treated as separate species but have also been considered as one, though possibly as subspecies. Together their ranges add up to an almost complete circumpolar distribution, without any apparent overlap, a strong pointer to their single species status. The Purple Sandpiper has the distinction of being the most northerly regular wintering wader. The birds travel south only just as far as they have to, and can be found in winter within their breeding range in Greenland and Iceland, Alaska, and many of the islands in the Bering Sea. At the southern end they reach the British Isles, Maryland, and California. It is not known how far individual birds will travel but it is possible that those in the southern part of the breeding range are sedentary, while the birds wintering further south have come from points further to the north.

The majority of arctic birds are comparatively tame during the breeding season, and may be approached closely. This is not only true of the nesting bird, which will sit tightly on the eggs in the hope that she will not be noticed, but of birds away from the nest too. The Purple Sandpiper is no exception but is also, and less usually, just as tame in winter. Then it is found in small flocks on rocky shores, fluttering among the seaweed-covered boulders, just escaping at the last moment as each wave threatens to engulf them. If approached slowly, they will not dash away but will carry on about their business, and it is not uncommon to be able to get within three or four yards before they will quietly move a little further off.

Their nests are on dry tundra, usually among mosses and lichens where they are perfectly camouflaged. Water must not be too far away as the young will be led to it for the rearing period. The birds breed in a scattered manner, probably defending a territory around the nest, but rarely breeding so densely that boundary disputes with neighbouring pairs become necessary; a walk across the tundra might encounter a pair every half mile or so. The incubating adult sits very tightly and flushing it is the only sure way of finding the nest, and attempting to follow a bird back to its eggs is almost certain to be unrewarding, not to mention frustrating. The nest itself is a much more pronounced cup than most other waders', perhaps four inches across and two deep. It is lined with dwarf willow leaves and other plant fragments, though how many of these are put there by the bird and how many merely blow in is not certain. The cups remain from year to year with remarkably little change—they probably fill with snow in the early winter which then freezes so that the shape is maintained until the spring thaw.

The clutch of four eggs is thought to be incubated solely by the male, but the female has been found incubating and both sexes apparently have brood patches, so that maybe the true story is not so simple. The male, too, is supposed to be wholly responsible for rearing the young. If disturbed off the nest, or when with the young, it has a most pronounced distraction display designed to lead a potential predator away from the eggs or chicks. Many waders, gamebirds, and waterfowl trail their wings along the ground in the so-called 'broken-wing' display, when the adult looks as if it is in dire distress and can easily be caught. The predator follows it only to be led a merry dance until it is a safe distance from the nest, whereupon the bird takes to the wing. In the Purple Sandpiper the display is further refined as the bird lowers both wings to the ground, humps its back, fans its tail, and then runs, quite fast, along the ground. It then looks extremely like a small rat or other rodent: indeed the display is called the 'rodent run'. A dark line down the centre of the bird's back and tail is exposed by the trailing wings and closely simulates the long dark tail of a rat, while the trailing wings represent the animal's feet.

Although the female apparently takes no part in the incubation or rearing, she does not leave the breeding grounds entirely after laying the clutch but moves to the coast and joins other females in flocks, feeding on the insect life on the shore or among the rotting seaweed of the storm beach. They thus do not compete for food with the growing young, which remain round the inland fresh water pools and streams. There is no apparent difference in the timing of the migration of the adults and young, though whether family parties actually stay together is not known.

Dunlin [Red-backed Sandpiper]
Calidris alpina

Dunlin are the commonest wintering wader in northwest Europe, and quite plentiful also in North America. Although they occur widely in the European arctic they also breed around the Baltic and in parts of northern Britain, suggesting either that they have only recently spread north, or that the more southerly breeders are a relic from the time when the whole population bred just to the south of the ice during the last ice age. The latter hypothesis is more likely as some differentiation into races has taken place. The high arctic race in East Greenland (plus small numbers in Spitsbergen) is almost exactly the same size as the low arctic and temperate race, but has a shorter bill, thus obeying the rule that more northerly breeders are likely to have smaller extremities.

They breed on low marshes and tundra flats, usually concealing their nest in a clump of vegetation. In some areas the nests themselves are on the drier slopes surrounding the marshes, to which the young are led as soon as they hatch. Each pair establishes and defends a self-contained territory which is maintained until the eggs hatch. In high latitudes these are four or five times larger than in regions further south, directly related with the more rigorous conditions of the arctic environment and with the variability of the predominantly insect food supply. Like other waders, they will eat plant food when they have to, but prefer insects if possible.

Thus the main function of the territories is to disperse the population in relation to the available food supply, but this is only necessary during the first part of the breeding cycle for, when the young hatch, the broods are taken to the nearest marshy areas. By this time the insect food is normally sufficiently abundant for there to be enough for all, without the need to expend time and valuable energy in defending an area. Both parents share in the incubation and rearing, although studies of non-arctic breeding Dunlin in Finland showed that the female only helped with the rearing for the first six days, leaving the male to carry on until fledging at about

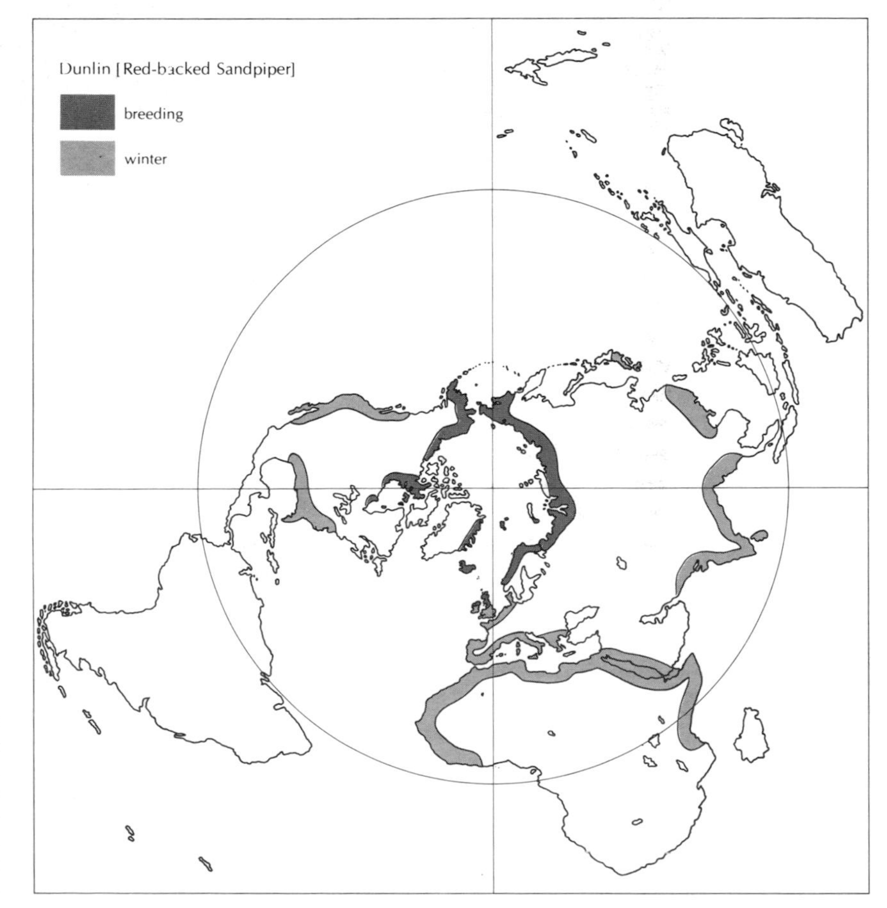

The breeding and winter ranges of the Dunlin, known in America as the Red-backed Sandpiper.

Opposite: The Hudsonian Godwit selects a dry hummock in marshy tundra for its nest. It breeds in the Canadian arctic and winters in South America.

The role of the sexes is reversed in the Red-necked Phalarope [Northern Phalarope]. The male has duller plumage than the female and he alone incubates the eggs and rears the young.

nineteen days. The normal clutch is four eggs and the incubation period twenty days. Unusually in arctic waders it is possible for a second clutch to be laid if the first is lost, but not if incubation is well advanced. This ability to relay is probably confined to the low arctic.

Dunlin feed largely on the larvae and adults of chironomids and tipulids, both common arctic insect groups. The adults eat larvae in the early and late summer, but take more adult insects in the middle period. This is when the young are being reared and they too take mainly adult insects from the ground surface. Thus there is some overlap but only at the period of maximum abundance. When the young have fledged they leave the breeding areas and flock to the coast, where they feed on shore and seaweed insects before departing on migration. Unlike other arctic waders most adults do not leave well before the young. These Dunlin will thus be the only adult waders of any species to stay on the inland tundra marshes after about the middle of August, and are consequently free of competition for the remaining insect food. This is of great importance to them because, again unlike other waders, they proceed with their annual wing moult before departing on migration. Their body feathers actually start moulting much earlier, often while still incubating, but the changing of the flight feathers is left until after the young have fledged, and in some populations may not be completed until after the migration. However in Alaska, for example, the entire moult is carried out before the adults finally depart from the breeding grounds in September. By this time the young, too, have left, travelling separately from the coastal regions to which they had gone.

In winter, Dunlin occur in very large flocks, often of 10,000 or more. They are typical birds of estuarine mudflats, spreading out at low tide to feed on small animals just below the mud surface, and gathering in tight packs to sit out the high tide on a traditional roosting site of a shingle bank or saltmarsh.

Stilt Sandpiper

Micropalma himantopus

This wader has a very restricted nesting range, both local and discontinuous. It breeds in the far northeastern part of Alaska and in low arctic Canada, on the mainland coast, southern Victoria Island, and in a few localities on the western and southern coasts of Hudson Bay. The autumn migration takes it through continental North America to winter on the eastern coasts of South America, and it does not occur on the Pacific coast at all. The Stilt Sandpiper is a typical wetland wader, rarely being found far from water both in summer and in winter, particularly the muddy edges of shallow pools and in marshy

Dunlin [Red-backed Sandpipers] defend a territory round their nest which varies in size according to the availability of food for the breeding pair.

ground. The birds arrive on the breeding grounds in late May and early June, the males a few days ahead of the females in order to take up a territory before the females join them. It has been shown that there is considerable fidelity by both sexes to both the territory and the mate of the previous year. Indeed they may even reuse the nest itself from the previous year, a characteristic that has been proved for very few wader species though it is common enough among, for example, the geese. However unlike the geese the birds do not travel and arrive as a pair but reform the pair bond on reaching the breeding places.

The territory is vigorously defended during courtship and laying, but the defence wanes during incubation when the need to preserve a food supply close to the nest decreases. In fact the off-duty adult may feed up to five miles from the nest, suggesting that the territory is more important in enabling the pair to reunite than it is in protecting an adequate food supply. Both parents incubate the usual four eggs, and detailed observations have shown that the males tend to sit during the day, from early morning to late afternoon, and the females during the night. The chicks hatch after about twenty days and are led as soon as they are dry from the comparatively dry nesting site to the nearby water. The female only stays with the brood for the first week, while the male carries on looking after them until the end of their three-week fledging period.

Buff-breasted Sandpiper
Tryngites subruficollis
The Buff-breasted Sandpiper has a comparatively small population breeding in rather few areas of arctic North America: northern Alaska, near Point Barrow, on the coast of the Mackenzie District of Canada, and on a few of the Canadian arctic islands. It migrates through the interior of North America to winter in central Argentina. The breeding habits of this bird are only imperfectly known, but it seems certain that in many instances the male is polygamous, courting and mating with more than one female, each of which then makes a nest, lays in it, and performs all the duties of incubating and rearing. The male plays no part in this at all. It is therefore most likely that, as in other species where the males does little or none of the work, he adopts a territory on arrival to which he attracts his future mate or mates. The nests are sited on sloping drier tundra and on the low ridges and drier banks beside streams, and are placed without regard to the territory boundaries. Like the male Pectoral Sandpiper, the male of this species is nearly ten per cent larger than the female, an unusual occurrence in waders but one that has the advantage of reducing food competition between them.

Although the breeding habits of the Buff-breasted Sandpiper are not fully known it seems probable that most males are polygamous.

Turnstone [Ruddy Turnstone]
Arenaria interpres
One of the most northerly breeding birds, the Turnstone occurs even in the far north of Greenland, north of 83°N and less than 400 miles from the North Pole. Its range is circumpolar and almost entirely within the high arctic. A few pairs still breed on islands in the Baltic, but these are most likely a relic

population from the immediate post-glacial period. This likelihood is strengthened by the fact that they are decreasing in number, although they are able to lay a second clutch if they lose their first and so take advantage of the much longer summer, unlike those breeding in the arctic.

There are two races of Turnstone. One occurs from northeast Canada, mainly Ellesmere Island, eastwards through Greenland, northern Scandinavia, and right across the top of Eurasia to Alaska. This is the typical race and it is blacker and less russet than the other, the Ruddy Turnstone, which occupies the rest of the breeding range within arctic Canada. The Ellesmere Island Turnstones migrate to Europe and join with the Greenland and European breeders to winter from the British Isles south to West Africa. A very occasional Ruddy Turnstone has been recorded there too, but their normal route is south through the continent of North America to winter in the southern United States south to Chile and Brazil. There is some evidence that the older birds winter further north than the young ones which do not return north to the breeding grounds in their first summer. The Alaskan and Siberian birds winter on the coasts of China, southeast Asia and Australia, and they also occur on the Hawaiian Islands in the central Pacific.

The birds nest mostly on the coast, sticking to stony beaches and gravel ridges, occasionally nesting in damper tundra but then on a dry hummock standing above the general marsh level. They arrive on the breeding grounds already paired, though the flocks may not split up for a few days. At this time insects are scarce and the birds continue the feeding habit which has earned them their name—that of flipping over small stones or flat pieces of dried mud in search of hibernating insects beneath. They are remarkably catholic in their tastes and will come close to buildings in the arctic picking up scraps and scavenging on rubbish tips. In winter, too, their diet is extremely diverse, even to the point of eating carrion from the bodies of animals and birds washed up on the seashore. The nest, though placed on dry ground, is never far from a stream or marsh to which the young will be taken. Both parents incubate the four eggs, for about twenty-one or twenty-two days, although the female is much more regular than the male who tends to be somewhat sporadic. However he makes up for this in completing the rearing of the young after the female has left on migration when the chicks are about half-grown. As with the Stilt Sandpiper it has been demonstrated that the chicks hatch at a time when the insect life is at its peak, and also that the adults' departure as soon as the young have fledged removes excessive pressure on the late summer food supply.

Turnstones do not moult until they reach their winter quarters. Here they live in rather small flocks, preferring rocky and stony areas to mud or sand. Immature birds

Turnstones [Ruddy Turnstones] winter on rocky coasts and get their name from their habit of flipping over stones in search of insects.

remain on the wintering grounds throughout their first summer but will often moult into breeding plumage, which is unusual for small waders.

Great Knot
Calidris tenuirostris
Very little is known about this northern wader. Its breeding range has not been clearly defined, but includes parts of northeastern Siberia. The very few breeding records all come from high ground, 1,000 to 2,000 feet, in mountain ranges around the upper reaches of the Kolyma river and eastwards into the Chukotski peninsula. Only a handful of nests have ever been seen, and these were placed on barren areas of mountain tundra with only scanty moss and lichen vegetation. The clutch size is the normal four eggs, and although there is no record of how the parents share the incubation, only the male has been found with the brood. The birds migrate down the Pacific coast of Asia to wintering areas in southeast Asia, Indonesia, and northern Australia.

Western Sandpiper
Calidris mauri
The Western Sandpiper only just penetrates into the arctic, being mainly confined to the subarctic of western Alaska, but extending to the north coast in a few areas. It has clearly been unable to adapt to the more rigorous conditions to the north but has become one of the most successful waders in its more usual breeding areas. Here it has stayed within a

Both parents of the Western Sandpiper share the incubation and rearing of the young.

comparatively small range, within which it occurs in relatively high densities, as many as 250 pairs per 100 acres being reported from its most favoured habitat of hummocky ground surrounded by rich marshland. In the high arctic it attains nothing like these densities. Both parents incubate and rear the young thus providing them with much better protection. However such behaviour is more suitable for the low arctic with its better food supply and may have helped prevent further colonisation of the high arctic. These birds migrate down the Pacific coast of North America to winter from California, round the Gulf coasts of the United States, and south to Peru.

Little Stint
Calidris minuta
With an overall length of less than a House Sparrow, this is one of the smallest of waders. Yet it migrates just as great distances as many of its larger relatives. The breeding range lies wholly within Russian Siberia, from close to the Norwegian border eastwards to the New Siberian Islands, while the habitat includes both the arctic tundra and the scrub areas to the south. Their main wintering area is in Africa from south of the Sahara right down to the southern tip, but some birds stay further north in the Mediterranean region and in southern Iran and India.

Little Stints place their nests in wet tundra or in marshy scrub, though always on a dry hummock. Some birds, however, prefer higher, drier areas away from the bogs. The usual clutch of four eggs is incubated by both parents, and they also take equal shares in rearing the young. Insects and larvae are their main food but in the absence of these they will eat plant seeds. The adults leave the breeding grounds in late July when the young have fledged, and complete their migration before they moult their wing feathers, a long process which is not finished until December. An interesting and fairly recent discovery is that immature waders of many species, including the Little Stint, also moult their wing feathers on the wintering grounds, starting in December and completing the change before the spring migration in April and May. This is presumably necessary because the first set of wing feathers is not sufficiently durable for three long migrations—south in the first autumn, north the following spring, and then south again the next autumn.

Red-necked Stint
Calidris ruficollis
A close relative of the Little Stint, the Red-necked Stint may yet prove to be only a subspecies. The breeding range is situated in northeast Siberia with an isolated locality on the Taymyr peninsula. In winter it is found throughout much of southeast Asia and in Australia and New Zealand. Its breeding habits have been little studied but seem similar to the Little Stint's, with the nest placed on dry hummocks in wet tundra, though as yet there are no observations on the share of the sexes in incubation or the rearing of the young.

Temminck's Stint
Calidris temminckii
When discussing the breeding habits of the Sanderling it was mentioned that Temminck's Stint was the only other wader known in which the female laid two clutches of eggs, with one looked after by each parent. This fact was discovered in Finland where the species breeds quite commonly, as it does from northern Scandinavia right across Eurasia to the Pacific coast. Only in northeast Siberia does it penetrate into the arctic; for the rest of the range it is to the south in the scrub and boreal zones. In winter, they are usually solitary birds, inhabiting freshwater marshes and pools well inland in Africa and southern Asia.

Temminck's Stints breed in what amount to loose colonies, although these may be no more than dense gatherings in suitable habitat of between twenty and forty pairs. Their preferred habitat is wet marsh with dry hummocks on which to place the nest. The normal breeding pattern seems to be for the pair to share incubation and for the male to do most of the rearing but, as explained above, in one area of Finland it was found that the female laid a clutch on which the male immediately sat, whereupon she moved some distance away and laid a second clutch which she incubated. Each parent then separately reared their own brood. It is not known how common or widespread this habit is but clearly it is more sensible that the male should incubate the first clutch, than that the female return to it after laying the second—the pattern apparently adopted by the Sanderling.

Least Sandpiper
Calidris minutilla
One of the smaller members of the sandpiper family, this bird breeds right across Alaska and Canada mainly in the sub-arctic belt but penetrating the arctic in the Mackenzie district and around Hudson Bay. It seems probable that despite the many ways in which arctic waders have avoided direct competition, there is just no room for yet another species, and being very small the Least Sandpiper would be at the greatest disadvantage if it found itself in competition.

The birds inhabit marshes and wet bogs for breeding and the four eggs are incubated for about twenty days. It is reported that the

Temminck's Stints share with Sanderlings the ability to lay two clutches in a season. Each parent then assumes responsibility for one.

One of the smaller waders is the Least Sandpiper. It breeds in boggy areas of low arctic Canada.

male takes the major part in the incubation but both sexes probably help to rear the young. They migrate south to winter from the southern United States to Peru and Brazil.

Sharp-tailed Sandpiper
Calidris acuminata
Almost nothing is known about this Siberian breeding species, concerning either its breeding habits or even the full extent of its range. It is said to nest along the north coast of eastern Siberia and to winter in southeast Asia, northern Australia, and New Zealand. It frequents grassy tundra in the breeding season but the nest, eggs, or young have apparently never been found.

Spoon-billed Sandpiper
Eurynorhynchus pygmeus
The extraordinary spatulate end to the bill of this species has long puzzled naturalists: no other wader needs such an extension with which to grasp its prey, and the birds do not seem to filter water through it either. Both mandibles have spoon-shaped ends, while the lower one has a shallow groove in which the tongue lies. Even the chick hatching from the egg already has the distinctive bill. Like the preceding species, the Spoon-billed Sandpiper breeds in Siberia, but in a very restricted area of the Chukotski peninsula, while it winters in Burma and other parts of southeast Asia. It nests in marshy tundra and the usual four eggs are incubated for about nineteen days. There are reports that the male does all the incubation but this unusual behaviour has yet to be confirmed. Particularly pleasing is its display flight, which is not common among high arctic waders as they usually have better things to do with their time and energy. The bird hovers 200 to 300 feet above the ground and then dives down, giving out a buzzing trill.

Medium and Large Waders

There are only nine species in this group of which four are primarily arctic in range (Ruff, Long-billed Dowitcher, Eskimo Curlew, and Whimbrel), while the other five are only secondarily so, but none of the first four penetrate so far as the preceding group of small waders. Their basic breeding biology is similar to their smaller relatives, and the role of the sexes in incubation and rearing of the young is just as varied. It is tempting to speculate that although even a large wader is a good deal smaller than say a goose or diver (which have successfully adapted to breeding in the high arctic), its essential body structure, including the long bill and very long legs, cannot be adequately modified to allow it to breed in the far north.

Ruff

Philomachus pugnax

The breeding plumages of the male Ruff are among the more extraordinary manifestations of the avian world. Many birds adopt a bright breeding plumage in order to attract a mate but this bird is almost without parallel in its very great individual variety. No less than thirty-eight different types have been identified, while minor differences on these basic types mean that in fact they are almost infinite in number. The winter plumage of both the male and female is a rather drab brown, spotted and streaked to produce what might be called a typical wader patterning. In spring the males moult into their bizarre breeding dress, growing long plumes round the head and neck that form eartufts and the 'ruff' that gives the species its name. The plumes are usually black, brown, rufous, or white, or some shade of these, with the ear-tufts always differing in colour from the ruff, so that a bird may have a black ruff and brown eartufts, or white and brown, or black

The Ruff's breeding and winter ranges.

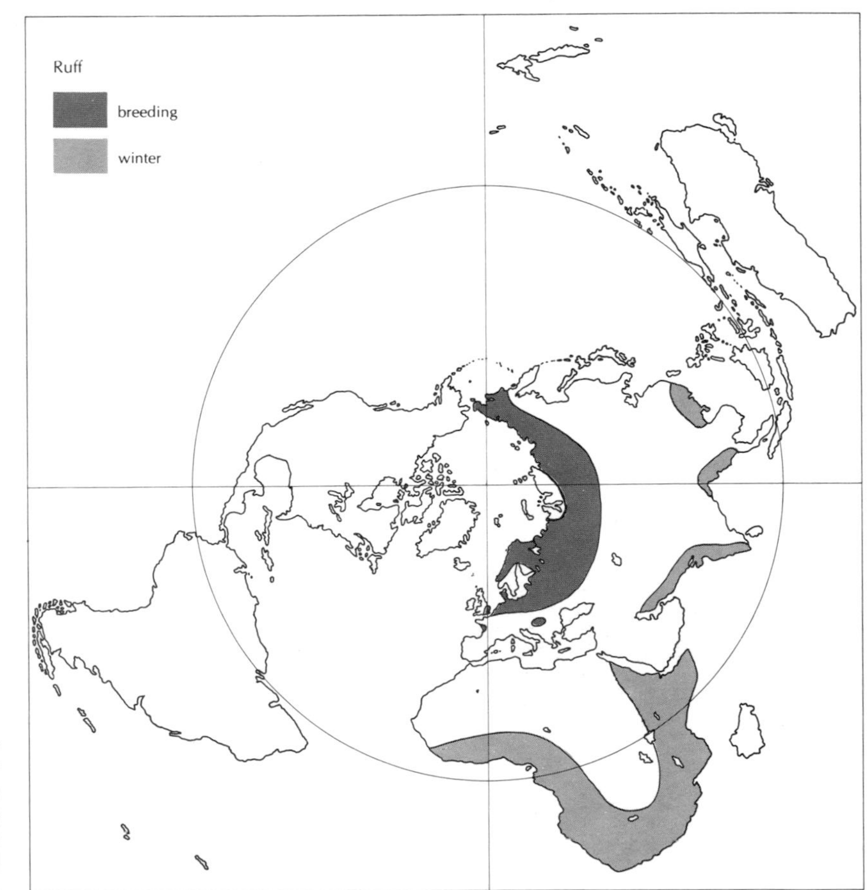

Opposite: Two male Ruffs displaying in their extraordinarily bizarre and varied breeding dress. A successful male may mate with several females.

and white, or indeed almost any combination, though some are commoner than others. Coupled with this plumage variation are differences in the colour of the bill and legs, which may be black, brown, yellow, red, orange, or green, and not necessarily the same in each bird.

The purpose of the elaborate plumage is, naturally, to attract a mate, but this is not done in the conventional manner of the male taking up a breeding territory and there displaying to draw the attention of a passing female. Instead all the males over quite a large area gather on a central display ground called a 'lek'. Here up to thirty birds form an approximate circle perhaps thirty yards in diameter, each male taking up position on a tiny territory only a few feet across, which he defends aginst the other males. They then display and posture, erecting their elaborate eartufts and ruffs to show them off to maximum effect, and flapping their wings and running and jumping. The females meanwhile gather on the outskirts of the lek and are attracted to a mate by his displays. No pair bond is formed; indeed a successful, dominant male may mate with more than one bird. When this has taken place the female moves away to make a nest and lay the eggs. The incubation and rearing of the young is solely her responsibility, for as soon as the mating period is over, the males begin to moult out

of their breeding plumage and move away to start their migration south, even though it may still only be late June or early July.

Ruffs breed from west Europe, including a handful of pairs in Britain, through northern Europe and Siberia. They make very long migrations: ringing has shown that birds from very far east in Siberia migrate in autumn through northwest Europe on their way to West Africa, where they winter in the Niger and Sénégal Deltas and inland at Lake Chad. Estimates of as high as half a million have been made for this last locality. The spring passage takes a different route, the birds heading northwest towards the Black and Caspian Seas, passing through southern Russia on their way back to the breeding grounds. Thus they travel in an enormous circle each year, in a course not dissimilar from that of the Pintail. Their winter habitat is shallow floodwater and estuaries.

Long-billed Dowitcher
Limnodromus scolopaceus
There are two species of Dowitcher, the Long-billed and the Short-billed. They are very similar in appearance and barely separable in the field. As their names imply, the chief distinction is in the length of the bill, but as there is a certain amount of overlap between the two, some people now consider them as subspecies. Contrary to expectation the larger, longer-billed Dowitcher occurs in the arctic, while the smaller, shorter-billed bird does not go so far north. The former breeds in the extreme northwest corner of Canada, in the Mackenzie District, in northern Alaska, and in northeastern Siberia. It is rarely found far from the coast and breeds in marshy ground with low vegetation.

The adults apparently arrive on the breeding grounds more or less simultaneously. The normal clutch of four eggs is incubated by both male and female but only the male rears the young, the female leaving on migration as soon as the eggs have hatched. Their wintering areas are in the southern part of the United States from California across to Florida, and south through Mexico into central America, where they live on fresh and salt water marshes.

Eskimo Curlew
Numenius borealis
For many years this species was thought to be extinct, with the last recorded specimen in Canada in 1932 and one seen in Texas in 1945. However there have recently been several sightings, mainly on passage through the southern United States or in the West Indies though confusion with other species makes its true status hard to assess. Formerly the birds were very abundant and huge flocks passed through the United States on migration. They were very easy to shoot, being tame and tending to pack in dense flocks, and even when a flock had been heavily shot at, it would apparently fly around in confused circles giving the gunners further opportunities. Hunting them is now of course strictly forbidden, but such protection was given literally at the last moment

and may yet prove to be too late. The Eskimo Curlew is very like the Whimbrel [Hudsonian Curlew] to look at, a fact which makes assessment of its true status more difficult, and really good views are necessary for certain identification. Its former breeding range and its breeding habits are not known, but it used to occur in northwest Canada and probably in northern Alaska, and may once have been much more widespread.

Whimbrel [Hudsonian Curlew]

Numenius phaeopus

There are three or four subspecies of Whimbrel that together achieve a discontinuous but nonetheless circumpolar distribution. The nominate race *phaeopus* breeds in northern Europe and western Siberia, migrating south to winter in both North and West Africa, right down to Cape Province, and also in the Middle East, India, and Ceylon. The birds breeding in Iceland and the Faroes are sometimes considered a separate subspecies *islandicus* but the differences are slight and they appear to use the same wintering areas in Africa, so probably may be considered as one with *phaeopus*. Going eastwards there is a wide gap in central arctic Siberia without Whimbrels, but then a large area of northeast Siberia is inhabited by the race *variegatus*. The wintering area for these birds is in southeast Asia and south to Australia and

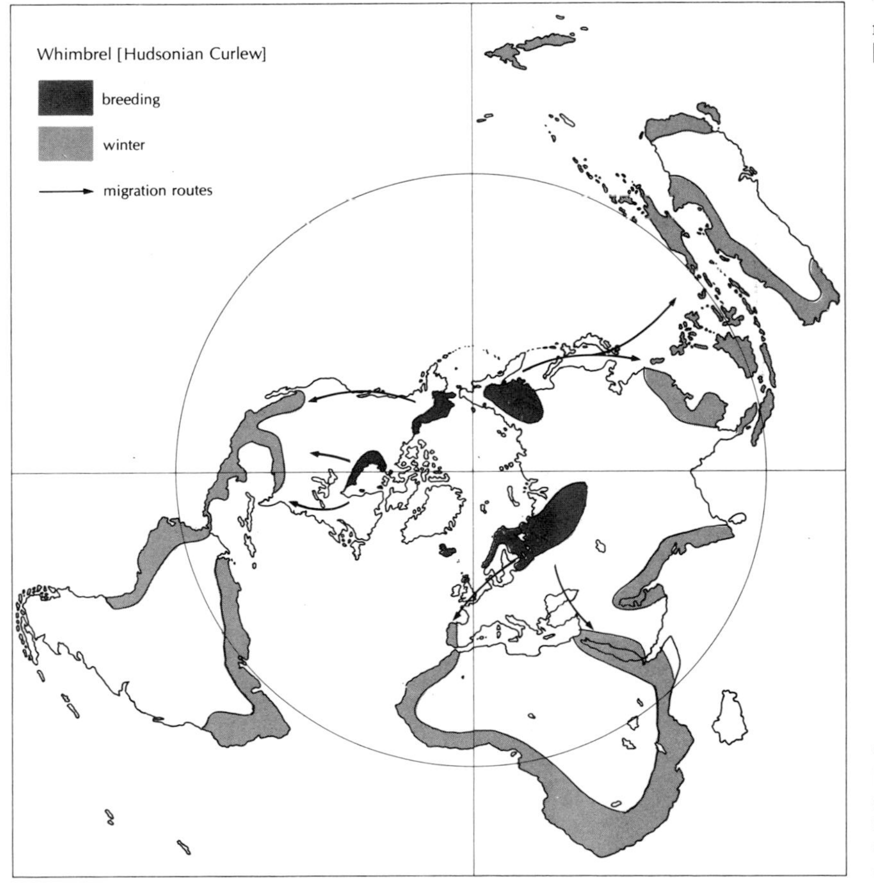

The breeding and winter ranges of the Whimbrel [Hudsonian Curlew].

Opposite: The Whimbrel [Hudsonian Curlew] gives out a long, bubbling mating call while flying on outstretched and quivering wings.

New Zealand. Finally there is a race *hudsonicus* which breeds in Alaska and northwest Canada, and around the south and west coasts of Hudson Bay. These birds winter in the southern United States and south to Chile and Brazil.

Whimbrel nest on wet tundra and marshy areas where there are dry hummocks for their nests, and they are well dispersed, holding large territories. Their breeding song is a long bubbling call, similar to part of the Curlew's song. Both parents share the incubation of the four eggs, and both also rear the chicks. However, as soon as they have fledged, the adults leave on migration while the young remain behind for a further two or three weeks. In the late summer both adults and young alter their diet from the usual animal food of insects and larvae, and eat largely berries of such plants as crowberry and cloudberry. The food value of berries is very high and well worth exploiting, even though the sight of one of these birds, with its long curved bill beautifully adapted for probing deep into soft mud, delicately picking off the berries and swallowing them might seem a little incongruous.

Hudsonian Godwit
Limosa haemastica
Bar-tailed Godwit
Limosa lapponica

Although these two species both reach the low arctic in parts of their range, in the main they breed in the subarctic in areas where there is scattered scrub and low trees. The Hudsonian Godwit is an uncommon bird, occurring in a few areas of south Alaska, northwest Canada, and Hudson Bay, breeding in the arctic in the Mackenzie District of Canada. Its wintering area is in southern South America. The Bar-tailed Godwit has a much wider range, breeding from northern Scandinavia across the full width of Siberia and into western Alaska. Birds from the west of the range winter in western Europe and North Africa, while the eastern population migrates to southeast Asia, Australia, and New Zealand. Both species lay four eggs in a typical wader scrape on the ground, commonly sited on a raised dry hummock in otherwise wet and marshy ground. The Hudsonian Godwit pair share in the incubation and rearing of the young, so far as is known, but the male Bar-tailed Godwit does the greater part of the incubation, although both parents rear the young. Unusually in such a long-legged water bird, it often perches in the tops of nearby low trees, keeping a look out for approaching danger.

Spotted Redshank
Tringa erythropus
Redshank
Tringa totanus
Lesser Yellowlegs
Tringa flavipes

The three 'shanks' occur mainly in the boreal or even temperate zones but all reach the low arctic in at least part of their ranges. The Spotted Redshank and Redshank are Eurasian species while the Lesser Yellowlegs is North American. All three are found by both fresh and salt water on migration and in winter.

The Spotted Redshank breeds in the arctic in northeast and northwest Siberia. Here it nests in boggy ground, but makes its scrape in a dry spot, lining it with willow leaves. The four eggs are incubated by both sexes, with the male apparently doing the larger share, and most if not all the rearing of the young. The females leave on migration first, and do not moult until they have gone at least part of the way to their winter quarters in the Mediterranean basin and Africa. Full-plumage adults, in their handsome sooty-black dress with white spots on the back and sides, arrive in Britain as early as July, where they moult into winter plumage. But before they have departed on their onward migration, more adults arrive in August, also in breeding plumage, so that birds in both winter and summer dress can be seen together. The first arrivals are presumed to be the females,

A Spotted Redshank on its nesting grounds. Birds in breeding plumage can be seen well to the south, having migrated before their autumn moult.

The Lesser Yellowlegs breeds mainly in the Canadian muskeg or scrub zone, but reaches the low arctic in a few areas.

the second the males, while the young birds do not get there until September. Very few actually winter in Britain; they merely use it as a convenient staging post.

Redshanks generally nest further south than Spotted Redshanks, with the exception of Iceland where there is a separate subspecies. The birds breeding in Britain are either sedentary or only move short distances to winter, but those occurring further north, in Iceland and Scandinavia, migrate south to winter in southern Europe and around the Mediterranean. Thus they have a 'leapfrog' migration which takes them right over both the summer and winter areas of others of the species. Their breeding biology is straightforward with both parents sharing the incubation and rearing.

The main habitat of the Lesser Yellowlegs is the muskeg or scrub areas to the south of the North American tundra. In northwest Canada and northern Alaska, however, it actually nests on the tundra. The breeding habits are similar to those of the Spotted Redshank, as both parents share the incubation but only the male looks after the young. In the autumn they migrate down the Atlantic coast or through the interior of North America to winter in Chile and Argentina.

Phalaropes

Phalaropes are highly specialised waders that spend more of their time swimming than walking or wading, and indeed are largely oceanic, living far from land throughout most of the year. Their feet have lobes along each of the toes to aid their swimming, and even small webs between the toes. Another adaptation for an aquatic life is the very dense plumage on the underparts This carries trapped within it a cushion of air that conserves warmth and helps the bird to ride high in the water. The principal feeding technique is to spin round and round on the water, darting and pecking with its bill at minute particles of food, including plankton, small crustacea, and insect larvae. The spinning operation helps to stir up the water and, coupled with vigorous paddling of the feet, brings food items to the surface.

These birds are notable for their reversed sexual plumages, with the female having the bright breeding colours. She is also a little larger than the much duller male—he has to perform all the parental duties of incubation and rearing of the young, and so naturally he needs the better camouflage. Although Phalaropes do apparently form proper pairs in some areas, most recent studies have shown that the females hold a territory in which they display to attract a male. When copulation has taken place the male selects a nest site, makes a shallow cup, and the female lays a clutch of eggs. Then she may return to her territory to display and mate with a second male and lay another clutch. On the other hand, male birds have also been reported to breed twice with different females. It seems clear that the pair bond is in any case very short lived as the females leave the nesting area soon after completing the egg laying.

There are two arctic species of Phalarope, the Red-necked (Northern), and the Grey (Red). If the alternative names for the latter species in particular seem odd it is because the English name was given to it for its grey winter dress (the only one seen in Britain),

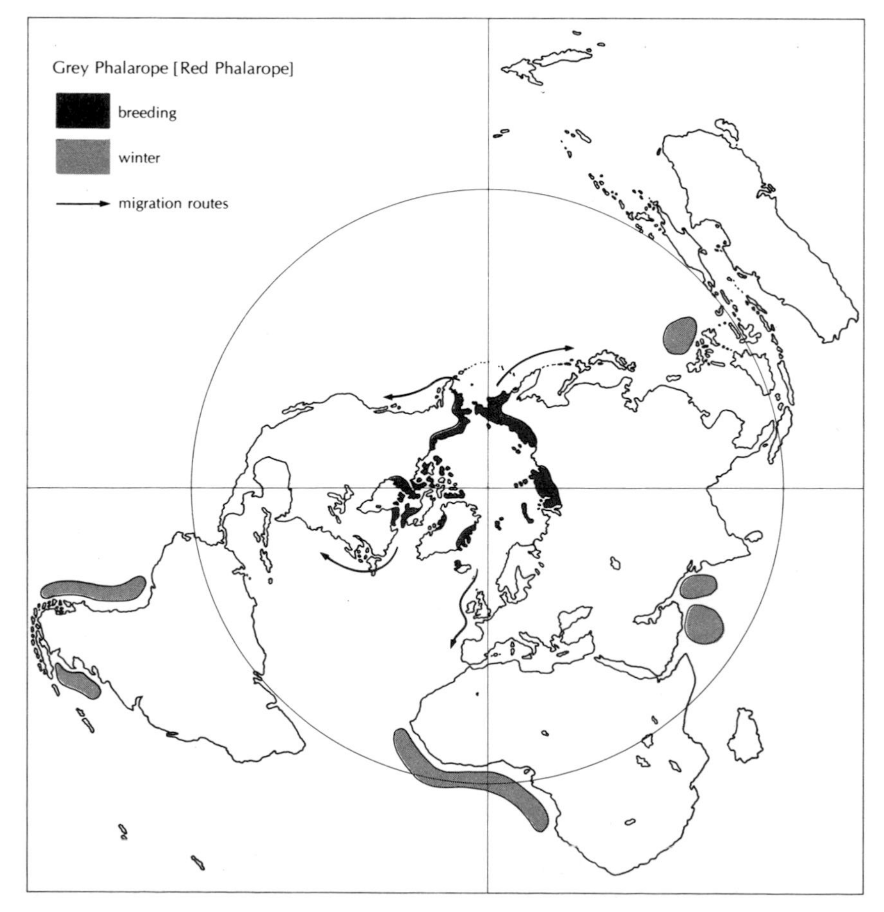

The breeding and principal wintering ranges of the Grey Phalarope [Red Phalarope].

while the American name refers to its gorgeous breeding plumage. The Grey Phalarope is essentially the high arctic species while the Red-necked is the low arctic one. However there is considerable overlap throughout the circumpolar range of both and it is something of a mystery to decide what it is that keeps them from competing. It may be that where they do occur together there is ample to eat, for there seems little difference in either their habitat or food requirements.

Red-necked Phalarope [Northern Phalarope]

Phalaropus lobatus

The Red-necked Phalarope occurs in the low arctic and in the boreal zone. A few pairs manage to breed in northern Scotland each year and there are small numbers in southern Finland, on the Baltic coast. They also nest well to the south of the tree line in northern Scandinavia, northeast Siberia, and Alaska. A small colony of these birds, consisting of four females and six males, was studied in northern Sweden. Two pairs appeared to form normally, if briefly, and each female laid a clutch of eggs on which their respective males sat. However the other two females each paired with two of the remaining four males. Both laid two clutches of four eggs only seven days apart, and each of their respective two males incubated them. Thus a situation where there was an excess of males was taken full advantage of to the benefit of the species as a whole.

The breeding habitat of these birds is freshwater pools and their well vegetated margins. The nest is nearly always placed in a clump of grass or some other plant, always well concealed, and takes the form of a small cup two or three inches across. After the female has laid her clutch, invariably four eggs, she leaves the area and the male gets on with the incubation for a period lasting about twenty days. At all times the Red-necked Phalarope is a very tame bird, or perhaps a better description would be indifferent to man. And this is never more true than when the male is incubating or looking after the young. It is possible to gently place the clutch of eggs in one's hand, resting this on the ground. The male, while this is done, may retreat a foot or two and watch anxiously. Then if one remains quite still, he will approach slowly, climb on the hand, and carry on incubating as if nothing had happened. I have done the same thing with a brood of very small young which were being brooded by the male. He did not actually climb on to my hand, but gently pecked it, as if in reproach, until his charges were released again. He then settled down on them right at my feet, not bothering to hurry them away but just content to have them snuggled under him again.

The male deserts the young as soon as they can fly, about three weeks after hatching. Both males and females appear to moult before they migrate, gathering in flocks on the sea. Their migrations are only partially known, becuase the main wintering areas are at sea, and observations from land give only an incomplete picture. There is certainly a route down the west coast of Europe and Africa, and autumn gales can often drive many birds inland, where they appear on lakes and reservoirs. This happens quite often with the Grey Phalarope in Britain. In North America there is an overland migration from central northern Canada to either the Pacific or Atlantic coasts, and each autumn in Eurasia there is a passage through the Caspian Sea of birds on their way to wintering areas south of the Arabian peninsula. Other wintering areas are thought to lie in the Atlantic off West Africa, in the Pacific off the west coast of Peru, and in the seas around southeast Asia.

Grey Phalarope [Red Phalarope]

Phalaropus fulicarius

The high arctic or Grey Phalarope, occurs much further north than the Red-necked, reaching Spitsbergen, northeast Greenland, and some of the high arctic islands of Canada. Its breeding habitat is similar, and the nest, too, is a small cup placed in vegetation, either a clump of grass or sometimes thick moss. In some areas it breeds in what amount to small colonies, with several females holding territories round one pool. Both they and the males perform display flights in flocks. Some researchers report a total absence of territorial behaviour while others are certain that it does occur, so clearly there are differences from one part of the range to another, a possible response to population size or availability of food.

A pair may copulate on land or on water, whereas the Red-necked Phalarope has only been seen to copulate on water, but the significance of this slight difference is not clear. The male incubates the four eggs for eighteen or nineteen days, a small but definite shortening compared with the Red-necked as befits a more northerly species. The fledging period has not been accurately measured. When they hatch the young are delightfully small and fluffy, their downy covering being a variegated pattern of black, white, and cinnamon. They can swim as soon as they leave the nest, within six to twelve hours of hatching, and once afloat they look exactly like floating bumblebees. They soon grow, however, and can fly within about three weeks.

The females leave the breeding area as soon as the male has begun incubation (or occasionally after the second mate has done so), while the males desert the young either just before or just after fledging. Both groups of adults apparently combine before migrating, having completed their moult into their dull winter plumage, and flocks of several thousands have been reported in coastal arctic waters in late summer. Their migration is almost entirely at sea, but gales cause 'wrecks' on both sides of the Atlantic, with birds blown well inland, particularly in September and October. There are three main wintering areas at sea—in the Atlantic off the west coast of Africa and off the west coast of South America, and in the Pacific off the northern coast of Chile—and each is known for its abundance of surface plankton in the winter months.

The Long-tailed Skua [Long-tailed Jaeger] is one of three skuas which breed in the arctic. They are predators of small mammals and birds, and chase seabirds to make them disgorge their food.

5 Seabirds

As their name implies, seabirds depend in part or in whole upon the sea. Some are only seen on land for the brief period of the breeding season, spending the rest of the year far out to sea. Others, particularly some of the gulls, have gradually adapted to man, his reservoirs and rubbish tips, and from being entirely marine and coastal birds are now found living far inland, but fish, plankton, molluscs, and crustaceans are the natural food of all the seabirds. Except for the adaptable gulls, all nest within short flying distance of the sea, often on cliffs whose feet are washed by the waves. Here there is maximum safety from land predators, and the easiest possible journey for their young, which in some species leave the nesting ledge before they can fly. Most seabirds are colonial, often in huge colonies with many species together on the same cliff. Small differences in nest requirement prevent too much competition so that, for example, all six species of European auks can be found together on Bear Island. The foods taken too vary slightly, although in the shallow arctic seas the prolific fish and plankton provide sufficient for all.

Skuas [Jaegers]

There are three species of skua breeding in the arctic: the Pomarine, the Arctic (Parasitic), and the Long-tailed. They are hunters, best known for their habit of pursuing other birds, particularly Kittiwakes and gulls, forcing them to disgorge the food that they are carrying in their crops, and then swooping down and catching it for themselves. They will often station themselves in small groups on the flight line between a Kittiwake colony and the sea. As the laden birds come flying back to the colony with food for their young, the skuas pounce and a spectacular aerial chase begins. The Kittiwake twists and turns, soars and dives, in an invariably vain effort to shake off its pursuers. It calls in agitation but eventually has to disgorge its load, and then, lightened, quickly flies away. The skuas swoop down on their ill-gotten meal, while the poor Kittiwake has to turn about and head out to sea once more, only to run the gauntlet again on its return.

All three skuas can be found breeding in the same area, so either their food requirements are sufficiently different for competition to be limited, or else in those areas there is enough to go round. In fact a study on the ecology of the three species in an area of northern Alaska found that there can be competition between them, but mainly when there is a food shortage. In these years some or even all of them might not breed, thus reducing conflict over the limited resources. The Pomarine Skua was found to be the most versatile feeder, taking mainly lemmings and the smaller microtine rodents when these were abundant, but also when necessary parasitising other birds. The Arctic Skua was exclusively a pirate, chasing Kittiwakes and Gulls, and thus only faced competition from the Pomarine Skua in years when lemmings and other rodents were scarce. The Long-

Only a handful of birds can winter in the arctic. The Black Guillemot does so where leads in the ice allow it to find food.

tailed Skua rarely chased birds, feeding principally on small microtine rodents. It thus overlapped with the Pomarine Skua in this requirement, and indeed the two species rarely bred near each other, and then only in years of maximum rodent abundance. All three also killed small birds or took eggs and young, but none relied on this diet.

Skuas nest on the ground, choosing a dry hummock in marshy tundra or else dry uplands. The usual clutch is two eggs, and both parents are assiduous in their care of eggs and young. A territory is defended round the nest and any animals, including man, venturing inside risks attack from one or both parents. The method of attack is a blow on the head, the bird almost always coming from behind. At the last minute it either stabs with its bill or lowers its legs and hits with the feet. In one locality in northern Scotland where a long-term study of Arctic Skuas is taking place, such attacks are made all the more painful because each bird is marked with metal and plastic rings on its legs. In addition, skuas sometimes indulge in an elaborate distraction display. One or both adults tumble over the ground, looking exactly as if one of their wings is broken. The tail droops, the wings beat on the ground, and all the time the bird keeps up a mewing call. A fox will invariably try to catch this apparently injured bird, which skilfully leads it further and further away from the nest or chicks, only flying off when the danger has abated.

Pomarine Skua [Pomarine Jaeger]
Stercorarius pomarinus

This bird breeds right round the North Pole with the exception of gaps in northeastern Canada, northern and eastern Greenland, and Spitsbergen, but it is now just beginning to colonise the last locality. The Pomarine is the largest of the skuas and perhaps for this reason it has not penetrated to the extreme northern areas already mentioned, or those of Siberia. Its wintering areas are entirely oceanic. It is found in the tropical and near-tropical waters of the Atlantic and Pacific where upwellings of cold water produce great richness of fish and plankton, principally off the west coast of Africa, in the Caribbean, and off the east coast of Australia. Very large numbers of other pelagic birds gather in such places, including shearwaters and tropicbirds, and the Pomarine Skua either parasitises these or catches its own food from the surface of the sea. It is seen passing European and North American coasts on spring and autumn migration.

These birds are dimorphic, that is they have two colour phases, a dark and a light. The most obvious difference is in the colour of the underparts—pale brown in the dark phase and almost pure white in the light. The incidence of dark and light phases has not been much studied in this species but it seems that there are more light phase birds further north, which would fit in with the heat conservation benefit that light plumage gives.

Pomarine Skuas are solitary nesters and defend a large territory. The clutch of two eggs is incubated by both parents for four weeks and the chicks are down-covered on hatching. They leave the nest within a few days and move short distances only, but in different directions so that they may be thirty or forty yards apart, an obvious defence against predators. The fledging period is between five and six weeks, though the parents continue to feed the young for a few more weeks after that.

Arctic Skua [Parasitic Jaeger]
Stercorarius parasiticus

The Arctic Skua is intermediate in size between the Pomarine and the Long-tailed. It has the widest distribution, with no gap in its circumpolar range, and penetrates both further north, to the most northerly parts of arctic Canada and Siberia, as well as further south, to northern Britain and round Hudson Bay, than the other two species. Like the Pomarine it winters mainly in the southern hemisphere in areas of the ocean rich in food.

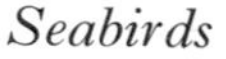

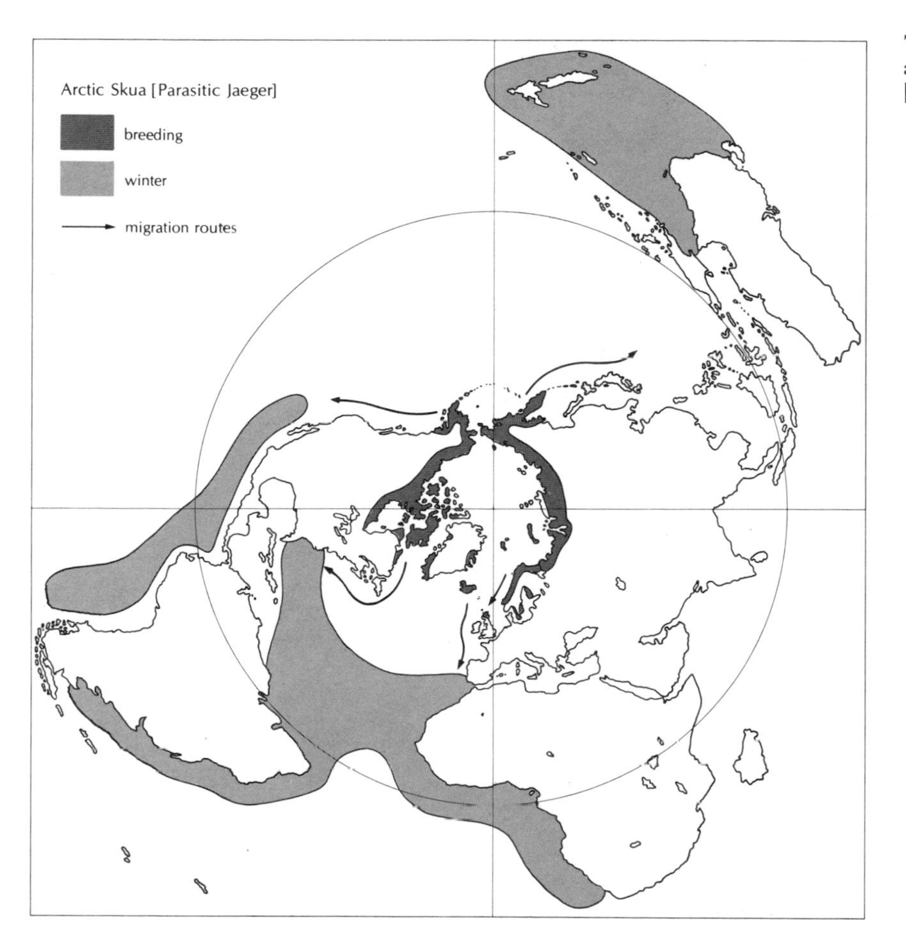

The breeding and winter areas of the Arctic Skua [Parasitic Jaeger].

The dimorphism of this bird is very pronounced. The dark phase adult is dark-brown all over, sometimes with slightly paler cheeks, whereas light-phase bird is pure white underneath, dark brown above. Intermediate stages also occur but are not common. The incidence of the two phases has been recorded in a number of localities: for example in northern Scotland dark phase birds make up about 75% of the population, and pale phase only 25%; in Iceland the proportions are 60% and 40%; while in the north of Greenland and in Spitsbergen, the dark phase only accounts for a mere 1%. In North America, too, the dark phase proportion decreases as one moves north. This is a clear example of white plumage being favoured the further north the bird lives.

Arctic Skuas breed in colonies, though within the colony each pair defends a territory round its nest. When the adults arrive in spring, they make for last year's territory where they meet their respective mates, probably for the first time since the end of the previous season. The incubation period is similar to the Pomarine Skua's at about four weeks but the fledging period is at least a week shorter, enabling the species to breed further north where the summer is shorter. During the incubation period especially, the off duty birds fly round in flocks, indulging in what can only be described as 'play', though there

A light phase Arctic Skua [Parasitic Jaeger] landing beside its nest. Each pair defends a territory against other pairs within a breeding colony.

may be some social significance attached to it. They plunge and dive through the air, twisting and turning, and chasing each other, all the time keeping up continuous screaming and calling.

Their pirating technique of obtaining food is mentioned above. They are also the species most adept at chasing and actually killing other birds. This is often done by a group, but a single Arctic Skua can usually outfly a wader such as a Sanderling with ease. The normal method is to strike it with the wings and feet until it is forced to the ground where it can be pounced on and killed, but occasionally the skua will actually catch the bird in mid-air with its beak.

Long-tailed Skua [Long-tailed Jaeger]
Stercorarius longicaudus

The smallest of the three skuas, this bird also has a circumpolar, high arctic distribution, extending south into the subarctic only in eastern Siberia and down the mountain ridge of central Norway. It is rare in Spitsbergen and some of the northern Canadian and Siberian Islands. Like the other species the wintering areas are the southern oceans, with one noted concentration off the coasts of Chile.

Long-tailed Skuas hold territories but do not breed in colonies. Their food largely consists of lemmings and small rodents but as these have cycles of abundance so the skuas have to vary their diet. In bad years they take small birds, particularly Snow Buntings, and the young of several species of

Long-tailed Skuas [Long-tailed Jaegers] are solitary breeders but still defend a territory round their nest, attacking anyone coming too close.

waders. They will also eat insects and feed these to their young. Their breeding is strongly correlated with the relative abundance of food: in good lemming years they lay two eggs, sometimes three, whereas in poor years most clutches will be of one egg, and many pairs may not lay at all. In addition the hatching and rearing success are equally dependent on food availability. The small size of this skua is an adaptation to high arctic breeding, allowing better use of small rodents, birds, and insects than the other two species could hope to achieve. In particular the feeding of the young on insects is important. The incubation period is twenty-three days and the fledging time of the chicks under four weeks—quicker even than the Arctic Skua, and not only allows breeding in the high arctic but gives the species up to an extra week to ten days in which to delay its breeding in a cold spring.

Large Gulls

Gulls are familiar seaside birds, now penetrating inland in many countries. In the arctic they remain coastal species but have nonetheless shown similar adaptations there in their ability to exploit the presence of man, as they have further south. Settlements in the arctic, whether towns, exploration camps, or meteorological stations, all have rubbish tips, and every tip soon has its flock of scavenging gulls. They are also scavengers in nature, eating carrion of all kinds, particularly along the shore or on the surface of the sea, so it is perhaps only to be expected that they should quickly find and exploit the waste products of man.

The relationships of some of the species of large gulls breeding in the northern hemisphere are rather tricky and still the subject of much discussion. It seems simplest to treat the Herring Gull as a single species that has a complete circumpolar range, by including in it all the many subspecies. (Some of these have been given full specific status by a few authorities, but the only one breeding in the arctic that seems to have achieved an accepted English name is Thayer's Gull of arctic Canada.) In addition to the Herring Gull there are three arctic white gulls, the Iceland, Glaucous, and Ivory, and finally there is the Great Black-backed Gull, which although strictly a temperate species, has begun to spread north in a few areas.

Gulls are mainly colonial breeders, nesting on cliffs and islands or even on flat ground in some areas. They lay two or three eggs in a shallow nest built from available materials such as seaweeds and grasses, and both parents incubate and feed the young during and after the fledging period. Although several of the species meet in winter at rubbish tips and following ships at sea, their summer foods and feeding methods differ.

Herring Gull

Larus argentatus

The most widespread of the gulls, this bird has a complete circumpolar range extending well into the arctic in most parts, to 80°N in Canada, but also coming south throughout most of the temperate zone. It is easily the most successful of the gulls as well, having no difficulty in adapting to new situations and foods. It breeds well inland, hundreds of miles from the sea in many areas, nesting on buildings in cities, and is colonising new sites all the time. If necessary, all or nearly all of its food can be found by scavenging on man's wastes. The bird nests in colonies, with the pairs often only feet apart. The two or three eggs are incubated for about four weeks and

A pair of Herring Gulls about to mate. This highly successful species has adapted to living in the arctic though originating much further south.

the young fledge in a further five to six weeks. They are fed by both parents bringing food in their crops. The young chicks stay in or near the nest for the first few days but after that may wander a short distance. However in a colony they run the risk of being attacked by another adult if they stray too near its nest, and the attack is quite likely to end in the chick being eaten. Apart from cannabilism, they are major predators of seabirds, taking the eggs and chicks of Guillemots and other species. The different races vary only slightly in their size and plumage colouring. The best defined subspecies in the arctic, Thayer's Gull, which breeds in the high arctic Canadian islands, is markedly paler than the Herring Gull, with much less black on its wing tips and a paler grey mantle and upper wings, as befits a bird breeding further to the north.

Iceland Gull
Larus glaucoides
The Iceland Gull has been considered conspecific with the Herring Gull; and there are also some people who think it is a separate species but believe that Thayer's Gull, here treated as a subspecies of the Herring Gull, is a subspecies of the Iceland Gull. It is a rare old confusion made worse because of lack of detailed knowledge of the ranges and therefore possible overlaps between some of the races and species, and because the whole complex of gull species is clearly in a state of flux, evolving all the time. The bird (as dealt with here) has a limited breeding range in east and west Greenland, and in a few areas in eastern arctic Canada. Despite its name it does not breed in Iceland, though it is a common winter visitor there. It winters either at sea or on the coasts of the northern

Several northern gulls have almost white plumage, though the immatures, like this Iceland Gull, are speckled with brown in their first two years.

Atlantic, reaching Britain and northeastern United States in small numbers. The birds breeding in Canada are sometimes called Kumlien's Gull and are given subspecific status.

Iceland Gulls breed in small colonies on cliffs or rocky islands. They tend to select sites in fjords and on sheltered coasts, rather than those facing the open sea, and nests have been found up to 2,500 feet above sea level. The nest itself is a typically bulky affair of grasses and seaweed perched on a ledge. Their normal clutch is two or three eggs, but the incubation and fledging periods do not appear to have been recorded. Although the bird will scavenge round rubbish tips, sewage outfalls, and fishing boats in the winter, it feeds on natural items in the summer, mainly diving for fish, an achievement shared with few other gulls. The bird flies slowly over the water and, on spotting a fish, goes into a quick dive with the wings half closed. It may submerge completely, but the dive is always a shallow one. Its diet also includes various shore-living animals such as crustaceans and molluscs.

Glaucous Gull
Larus hyperboreus
This species has a complete circumpolar range, extending through the entire arctic with the exception of a few Canadian islands. It effectively replaces the Great Black-backed Gull of the more southerly boreal and temperate zones, and some authorities treat it as a subspecies of the latter but they can co-exist without interbreeding. Like the Iceland Gull, it is almost pure white, lacking any black or dark grey in its mantle and upper wings and is thus truly adapted to the arctic. It does not

migrate far south in winter, staying in waters closest to the pack-ice or around the most northerly ice-free coasts. If it came any further south it would be in direct competition with the Great Blackback.

Glaucous Gulls are predators by nature, their prey being the enormous seabird colonies. Any Little Auk or Guillemot colony will have several of these gulls patrolling up and down, endlessly flying to and fro along the cliff or scree, eyes ever alert for the chance of snapping up an egg or a youngster. While this may be easy with Guillemots, which nest on ledges out in the open, the Little Auks breed deep in rock crevices and under boulders. Here the gull looks for an adult just landing or just emerging from its nest tunnel, and pounces to grab it before it can either get out of sight or launch itself safely into the air. Although the numbers of birds that each Glaucous Gull takes must run into many hundreds during a season, the seabird colonies are generally so vast that even predation at this rate makes no appreciable difference to them.

Another, more serious predation is on Eider colonies. These are almost invariably situated on small islands and rocky skerries, where the gulls too like to breed. Although the Eiders sit very tightly, the gulls are expert at finding the nests when the females are off. In Spitsbergen the Glaucous Gulls have been increasing in recent years and the Eiders decreasing, and it is tempting to link these

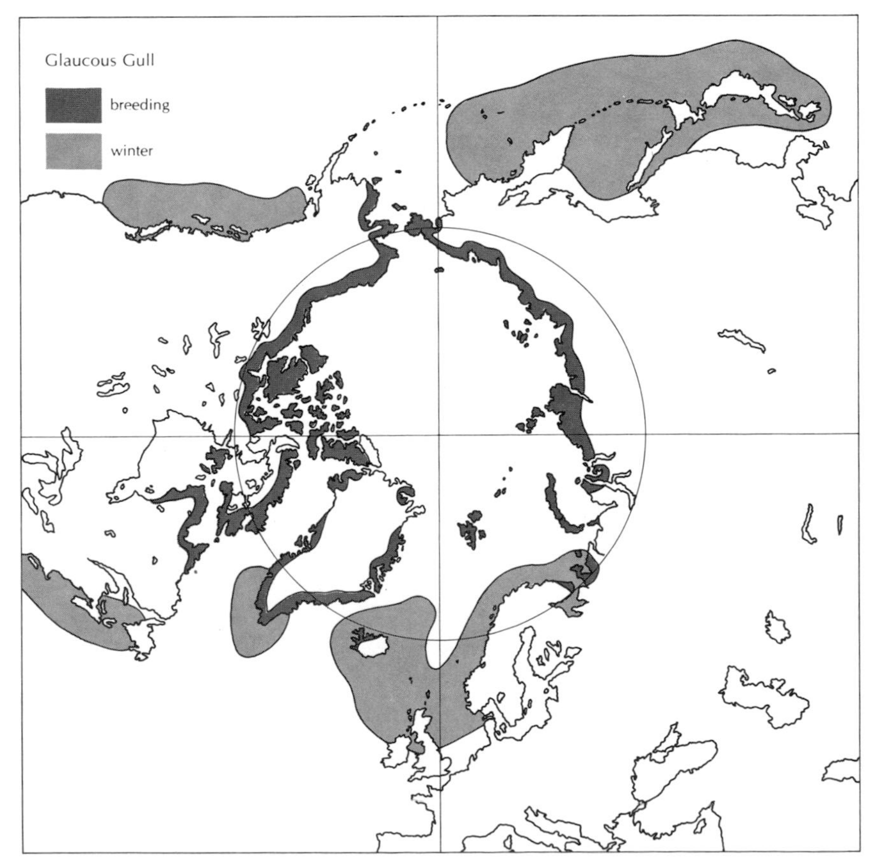

The Glaucous Gull's breeding and winter ranges.

The Glaucous Gull is one of the major predators on ground nesting birds in the arctic, especially the Eider Ducks.

two trends even though direct evidence is lacking. What is certain is that large numbers of gulls now winter there where formerly none could for lack of food—several hundreds live off the rubbish tip formed by the coal mining community in Isfjord. They not only survive the winter successfully, but are in good condition early in the season, with no distance to go to reach their nesting sites. Breeding may be in small colonies or in solitary pairs. The two or three eggs are incubated by both parents for about twenty-eight days, and the young fledge in five to six weeks.

Ivory Gull
Pagophila eburnea

The Ivory Gull is the most completely arctic of all the larger gulls, with a breeding distribution confined to high arctic islands and coasts. It nests on the Russian islands of Wrangel, New Siberia, Novaya Zemlya, and Franz Josef, on Spitsbergen, in northern Greenland, and in a handful of places in north Canada. In winter it stays among the pack-ice, rarely moving far, although stragglers do occasionally turn up well to the south. The bird breeds in small colonies on cliffs up to 1,000 feet high, but sometimes low islands are used where equal safety from arctic foxes can be assured. Its nest is a bulky affair of available vegetation. Normally the clutch is two, and the eggs are incubated for twenty-four or twenty-five days. Both parents take spells at this, which in one colony varied from a few minutes to eleven hours. The young fledge in rather over five weeks.

Two unusual sites for colonies have been discovered in widely separated parts of the arctic. In northern Canada a colony of between 75 and 100 pairs was found on a floating ice island 25 miles from the nearest land. There are a number of these ice islands in the arctic sea, some of them large anough to land an aircraft on, and they are thought to be very large icebergs calved from a glacier. This particular one was only a few hundred square yards in extent, and furthermore was covered in rocks and stones, which no doubt made it more attractive to the birds. The other unusual site was reported in southern Spitsbergen, not by an ornithologist but by a Polish glaciologist. He discovered a number of small colonies of Ivory Gulls breeding on the cliffs of nunataks, which are rocky out-crops and mountain tops sticking up out of the inland ice. None of the five colonies that he found was less than twelve miles from the nearest fjord, and some were as much as twenty miles or more from the open sea.

In summer Ivory Gulls feed on a wide variety of marine organisms, including small fish and crustaceans, which they take by hovering just above the surface of the sea, being reluctant to actually alight. However their main diet in winter is obtained by scavenging, and they will move in on a seal killed by a polar bear before the animal has barely had time to finish his meal, picking at the skin, bones, and offal. Their dietary habits become even less pleasant at times as they eat both these animals' droppings. Gulls also used to be common at the old arctic whaling stations, feeding on the offal and other waste products. By having these specialised feeding habits, these birds can comfortably survive through the winter, staying close to the edge of the arctic pack-ice or even wandering hundreds of miles over it looking for leads and therefore the likely presence of seals. They are further adapted to such living by having completely white plumage and a smaller bill and shorter legs than other gulls of their size.

Great Black-backed Gull

Larus marinus

The Great Black-backed Gull has a mostly boreal and temperate distribution on both sides of the Atlantic but has been spreading north into the low arctic in recent decades, and where it does so it appears to oust the Glaucous Gull. The best example of this has taken place in Iceland. Here the Great Black-back has long been a breeding species but in the last fifty years or so it has spread and increased, while the Glaucous Gull, formerly quite widespread over the whole country, has retreated before it and is now confined to the northwest peninsula. Climatic amelioration has allowed several more southerly species to colonise and spread through Iceland this century and has also contributed to the retreat of one or two more northerly species, so competition may not be the sole reason for the replacement of one gull by the other. In the last two decades these birds have reached Spitsbergen and now breed there in small numbers, and a similar picture is presented

Great Black-backed Gulls, related to Glaucous Gulls, are just as voracious predators on other birds, being capable of swallowing a Puffin whole.

in parts of the Canadian arctic where reports of them from Baffin Island and northern Hudson Bay are getting more common.

Small Gulls

The four species of small gulls form rather an amorphous group from four different genera. One of them, the Common Gull, only just reaches the arctic but the other three, Sabine's and Ross's Gull, and the Kittiwake, are proper arctic species. Not being very closely related (except in size) their habits have more differences than similarities, but like the larger gulls, they are colonial breeders, nesting on cliffs or small islands. The clutch size is two or three and both parents share in the incubation and rearing of the young.

Sabine's Gull

Xema sabini

Apart from three small areas on the north coast of Siberia this species is confined to Alaska, arctic Canada, and Greenland, and even here its distribution is discontinuous and patchy. It is a migrant and can occasionally be seen passing by the coasts of Europe or North America but its wintering areas have not so far been discovered; presumably they are well out to sea. Sabine's Gulls have been well likened to waders, with which they show several similarities both in their behaviour and in their breeding habits. They are small, dainty birds with quick feeding movements: they peck objects from the edge of pools, swim on the surface, spinning round like phalaropes, or run over mud feeding just as waders do. They also dive for fish, plunging clumsily into the water like heavily built terns.

They breed in scattered groups of up to twelve pairs that hardly warrant the term colony, but often their nests are among those of Arctic Terns from whom they no doubt gain protection against foxes and other predators. The nest is a shallow cup lined with grasses and seaweed, usually placed on the ground on a small island, sometimes offshore, or perhaps in a lake, but always near the coast. Indeed it is rare to find them breeding more than a quarter of a mile from the sea. The two or three eggs are incubated for between twenty-one and twenty-six days. Immediately on hatching and as soon as the chicks are dry, they are led from the nest by the parents to the edge of the nearest water. This is a very wader-like habit and is aimed at reducing the dangers of predation on the chicks and increasing the time they can spend feeding by taking them to the food supply instead of the other way round. In addition the parents have a rudimentary form of distraction display like many waders and like the skuas, but virtually unknown among gulls. The timing of the hatch of the young appears to be well correlated with the peak availability of insects, again similar to some wader species, for unlike any other gull, insects form the bulk of the diet of the young, at least while they are small.

Ross's Gull

Rhodostethia rosea

Ross's Gull is both one of the most attractive seabirds and, until comparatively recently, one of the most mysterious. Its plumage is suffused with a delicate pink giving it a beauty rarely attained by other gulls. The species was first discovered in 1823 by Sir James Clark Ross who was accompanying another famous polar explorer, W. E. Parry, in an attempt to find the North West passage. Ross collected two birds on the Melville Peninsula in arctic Canada. Four years later the two men again encountered the bird, this time a hundred miles north of Spitsbergen when they were trying to reach the North Pole. Scattered reports were received over the next fifty years, all from the polar basin, but no-one came any nearer to discovering where the species bred. Then in 1885 a nest was found in west Greenland, but this turned out to be a fluke as no other pair has since bred anywhere within thousands of miles of the country. Nansen saw several Ross's Gulls,

Sabine's Gulls lead their chicks, as soon as they are dry, to feeding areas in the nearest marsh; a very wader-like habit.

including small flocks, during his trek from the *Fram* towards the North Pole and then south to Franz Josef Land, but there was still no evidence of breeding. Most of the sightings, including Nansen's, were made in late summer and early winter, so it became clear that at this time the species was regular in small numbers in the pack ice north of Spitsbergen and some of the Siberian islands.

Then in 1905, to general astonishment, the breeding grounds were located, not on some high arctic island as all had supposed, not even on the tundra, but in the marshy and well-wooded valleys of a few eastern Siberian rivers. Here small colonies were found among willow scrub, making nests of grass and leaves, and each pair laying their three or four eggs in pleasant wooded, and certainly not arctic, surroundings. Their incubation period has been determined at twenty-three days, but the fledging period is not known. From

Although breeding in Siberia and wintering in the polar basin, Ross's Gulls occasionally appear far to the south, like this one photographed in Britain.

these breeding grounds, the Ross's Gull makes a now well-marked northward migration to the arctic pack-ice where it spends the winter searching for its fish and crustacean diet among any leads and open water it can find. At Point Barrow on the northern Alaskan coast a regular autumn migration can be seen in October and November of hundreds and thousands of the birds moving east into the polar sea. However no return passage occurs there in spring, so presumably the birds then take a different route.

Kittiwake
Rissa tridactyla
There are few noisier things in the bird world than a really large seabird colony, especially the shrill screaming of the Kittiwakes yelling their name 'kitti-waak kitti-waak' which echos and re-echos from the cliffs. These birds breed patchily right round the arctic but are absent from much of northern Siberia and many of the Canadian arctic islands, probably because of a lack of suitable breeding cliffs. They also nest quite far south in some regions, for example on Newfoundland, round the British Isles, and right round the Kamchatka Peninsula in eastern Russia. In winter they stay in oceanic waters, far from land.

Colonies of Kittiwakes can run into many thousands. The nest site is the merest ledge on a cliff on which the pair build a considerable structure. The base is usually seaweed which as it dries becomes cemented to the ledge, on top of this comes a layer of grass and mud, and it is completed with a grass lining. The finished nest may considerably overhang the original tiny ledge. This cliff habitat gives them complete safety from foxes but they suffer from a regular loss of eggs and young falling off the nests and being dashed

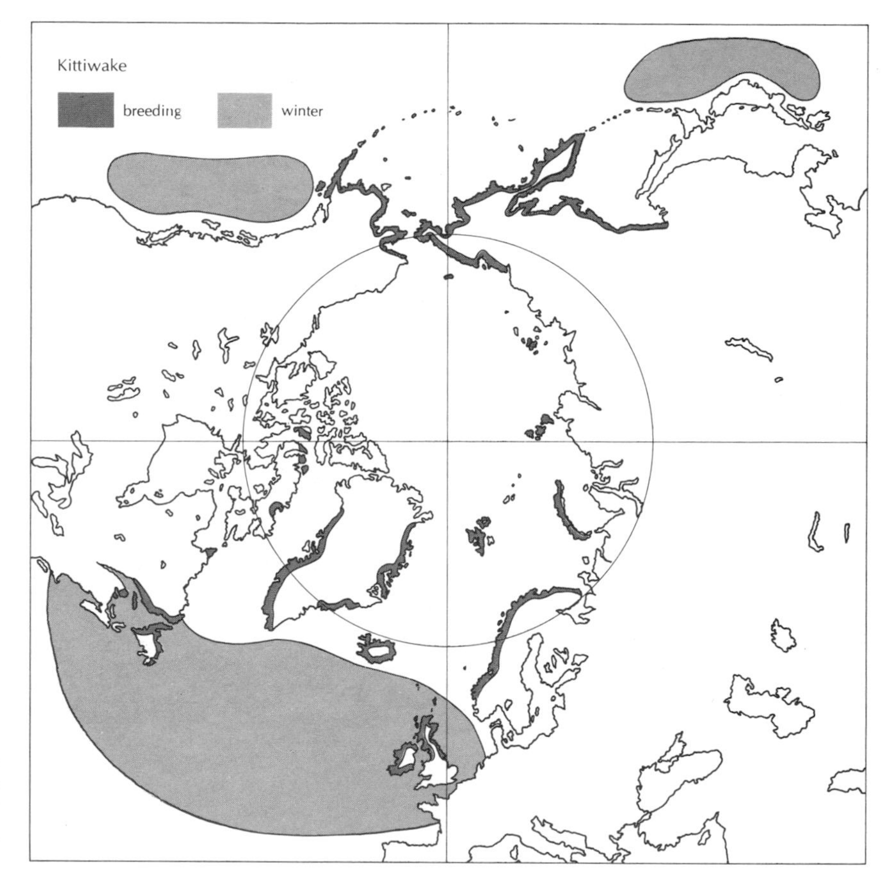

The breeding and winter ranges of the Kittiwake.

Opposite: Kittiwakes breed in dense colonies on cliffs, enlarging tiny ledges by building mud and seaweed nests.

to pieces. The normal two eggs are incubated in about twenty-five days by both parents taking two shifts a day each. The young fledge in a further seven weeks, and are fed on small fish and crustaceans which the adults take from the surface of the sea or in shallow dives. It is interesting that in the last fifty years Kittiwakes have taken to nesting on buildings in many areas, and are now quite common particularly in fishing ports around the North Sea. Window ledges are the favourite site.

Common Gull
Larus canus
The Common Gull breeds predominantly in the boreal region from northern Britain right across the USSR to the Pacific coast, and in the northwest of North America. Only in parts of Iceland, Siberia, Alaska, and Canada does it reach the tundras of the low arctic. It is a migratory species wintering along the coasts of western North America south to California, in Europe as far south as North Africa, and down the Asian coast to Indo-China. In many areas there has been an increasing tendency to winter inland, following rivers to new wintering sites on reservoirs. Some roosts total up to 50,000 birds which flight out each morning as much as twenty miles to feeding areas on farmland and rubbish tips. The usual nesting site is in scrub or thick heather, though tree nesting has been reported from many areas. In the arctic the sites are on the ground or in low scrub. The normal clutch is two or three eggs, incubated

by both parents for about three and a half weeks, and the young fledge in a further five weeks.

Terns

There is only one tern in the arctic, rightly named the Arctic Tern. These birds are close relatives of the gulls but are better fliers and divers, catching most of their food by plunging beneath the surface. They nest on the ground in colonies, and are assiduous in the defence of their eggs and young. Any predator is attacked vigorously, either by beating it about the head if it is an animal or by close pursuit and dive bombing in the case of a bird. This splendid aerial defence encourages other species less skilful at such tactics to nest within its colony, and is common among ducks, particularly the Eider and Long-tailed, some of the waders, and Sabine's Gulls.

Arctic Tern
Sterna paradisea

The Arctic Tern is one of the greatest travellers among birds. It breeds right round the arctic, extending south into the temperate regions round the coasts of northwest Europe and parts of North America, and it winters among the pack-ice of Antarctica. This involves a minimum journey of 10,000 miles, and some birds are known to travel much more. For instance the birds breeding in Newfoundland and Labrador, and probably

The Arctic Tern is a ferocious defender of its nest, attacking any animals including man that come too close. stabbing with its sharp bill.

Opposite: The Arctic Tern breeds throughout the arctic but winters in the antarctic, over 10,000 miles away. No other animal or birds sees as many hours daylight in a year.

also those in the eastern areas of the Canadian arctic, first cross the North Atlantic to the European side at the start of their autumn migration, and then fly south down the European and West African coasts. At the 'bulge' of West Africa they cross back to the South American side and complete their journey down that coast to Antarctica. This might seem an extraordinary route to take, and certainly not the straightest or most direct, but what they are doing is making use of the most favourable winds for as much of their journey as possible. The old sailing ships also used these trade winds, but Arctic Terns have known about them for thousands of years.

Once in the antarctic seas, the terns hunt along the edge of the pack ice seeking concentrations of fish and plankton. They may cover hundreds or thousands of miles during a winter before they turn north for the long return spring flight. It is of course summer in the south when they are there and it has been worked out that this bird sees more hours of daylight in a year than any other living bird or animal, spending the majority of its time in one or other area of midnight sun.

Its nest is a shallow scrape on the ground, lined with just a few small stones or bits of debris, and the two eggs are so perfectly camouflaged that they can hardly be seen from a few feet away. The incubation period is about three weeks and the young fledge in a further four.

Auks

Highly specialised marine birds, auks come to land only to breed. They are adapted to a life on and beneath the surface of the sea and, although they have not evolved so far as the penguins of the southern hemisphere and become flightless, they have nevertheless gone some way towards it. Their wings are relatively short and narrow and are perhaps better adapted for swimming underwater than

A delightful and familiar bird of temperate lands, the Common Puffin breeds in seabird colonies in Greenland, Iceland and Spitsbergen.

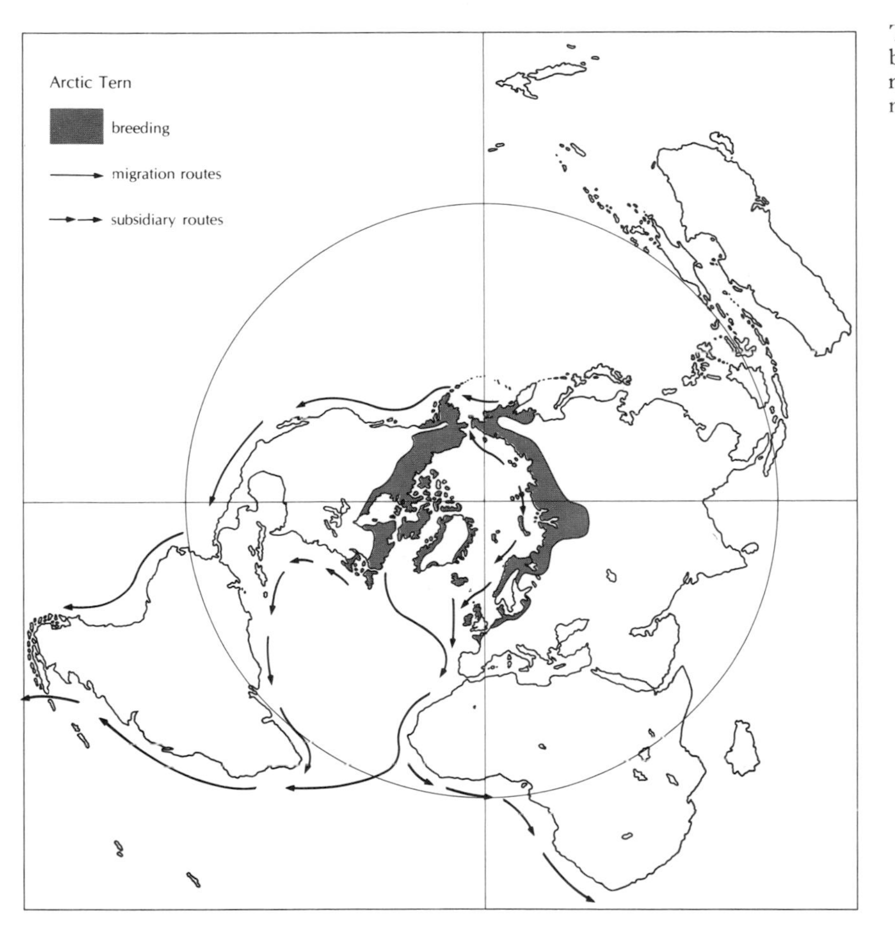

The Arctic Tern's breeding range, and its main and subsidiary migration routes.

for flying, and their legs are set far back for better propulsion. This makes movement on land difficult and they either stand almost upright or rest on their bellies. In addition, some of their migrations appear to be made almost entirely by swimming rather than flying, yet they can cover a thousand miles or more this way. Auks nest either on cliff ledges or in crevices and among boulders. They only lay one egg which both parents take turns in incubating. The young are usually taken to the sea when they are still quite small and before they can fly. They will however have already acquired a completely waterproof plumage. One exception is the Black Guillemot whose young are fully grown before they leave the nest. The other species of auk are the Common and Brunnich's Guillemots, and the Razorbill.

Common Guillemot [**Common** or **Pacific Murre**]
Uria aalge
The Common Guillemot occurs in the low arctic, and is replaced in the high arctic by the Brunnich's Guillemot. It breeds along the Norwegian coast, around Iceland, on Newfoundland, the west and south coasts of Alaska, the Kamchatka Peninsula, and on many of the islands in the Bering Sea. It is also found on coasts to the south (around the British Isles and north Spain) in the temperate

The close-packed ledges of a Common Guillemot [Common Murre] colony. Disturbance can cause great loss of eggs and chicks falling off the cliff.

zone. There is only slight overlap between the two species but where they do occur together they seem to live quite happily with no obvious competition, except possibly for nest ledges. The seabird colonies are, after all, located in areas of very abundant sea food which is not subject to the kinds of fluctuations that occur on land. Bad weather or the cycles of abundance of lemmings and rodents can lead to great variations in breeding performance of some birds, but such variation is hardly known for auks. Really bad weather can hamper them a little, but their cliff nesting ledges are unlikely to be seriously affected and their food supply hardly at all.

The birds incubate their single egg for about thirty to thirty-five days. The egg is extremely well adapted to life on a narrow ledge with no nest to hold it on. It has a pronounced point at one end and is rounded at the other, so if moved it will only roll in a very tight circle. Even so quite a number do fall off, particularly if the birds are scared and leave in a hurry. At such times, too, large gulls immediately take advantage of the parents' absence and swoop down to seize their booty. It is for this reason that human disturbance of a colony is so damaging. The young chick leaves the nest at only two or three weeks old when it is barely one-fifth the size of its parents and flightless, having spent most of its energy in growing a complete suit of waterproof plumage, though not its wing feathers. The parents escort it to the water's edge, a journey that may include scrambling over boulders on the beach below the cliff, and then swim with it out to sea. The advantage of this behaviour is that it enables the parents to take the young ones to the food instead of vice versa, a considerable saving in time and energy. In addition they can move to feeding areas much further from the colony than the parents could otherwise reach. Their principal food is fish, caught by underwater

A 'bridled' Common Guillemot [Common Murre] with its very pointed egg. The bridling increases from south to north in colonies in the Atlantic.

pursuit. The family parties gradually move away from the breeding coast and head towards the wintering areas out at sea, though they rarely go beyond the edge of the continental shelf.

Guillemots show an interesting plumage variation. A proportion of them have a white ring round the eye with a stripe running back from it. This 'bridled' form as it is called, increases as one goes north, from about 1% in colonies in the south of England to as much as 50% in Iceland and Bear Island. However it is completely absent in the Pacific race of the Guillemot. Its significance is not known though the increase of bridling from south to north is remarkably similar to the changing colour phases of the Arctic Skua and the Gyr Falcon, and it may yet be shown to have some function in relation with temperature, with which it certainly correlates.

Brunnich's Guillemot [Thick-billed Murre]

Uria lomvia

This species is very similar to the preceding one in size and shape but it shows one adaptation to a life further north in that its bill is stouter, being both shorter and thicker. Brunnich's Guillemot breeds on islands in the Bering Sea and on the coasts on either side, on the northern and eastern arctic islands of Canada, in Greenland, Spitsbergen, Iceland, and on most of the island groups off the north coast of Russia. It winters in the nearest open water to the breeding area. In the case of the west Greenland and eastern Canadian birds this means the Grand Banks off Newfoundland, whereas the Common Guillemots breeding in that area move south.

The birds breed in enormous colonies. Up to two million have been estimated for one site in West Greenland, while three sites in Canada are believed to hold over four and a half million birds between them. These

Brunnich's Guillemots [Thick-billed Murres] breed further north than Common Guillemots. Their stouter, thicker bill is an adaption to colder conditions.

colossal numbers naturally require vast quantities of food, and so all large colonies are found close to areas of the sea particularly rich in fish. The egg varies in background shade from pale green to bright blue, and has differing amounts of brown and black blodges on it. Its incubation period is about thirty-three days, and the chick leaves the nest in approximately three weeks. At that stage the young are unable to fly but, like the Common Guillemot, make their way to the sea as best they can. While this may be fine when the colony is on a sea cliff, there are some in Spitsbergen for example which are over a mile from the water, and it has still not been discovered how the chicks make their way over the rocky scree, tundra, and bog. It seems highly unlikely that they would in fact be able to travel that far or that the parents would be able to accompany them, and it may be that these birds stay on their ledges longer, at least until they are capable of sufficient flight to reach the sea.

The family party of Brunnich's Guillemots move steadily away from the colony, and as the young grow so the adults start their annual moult. They do not become completely flightless at this time, like the waterfowl, but there is no doubt that their already only just adequate flying ability is markedly impaired. This is no inconvenience, however, as they migrate to their wintering quarters largely by swimming. Journeys of well over a thousand miles are accomplished, with birds from Spitsbergen and the north Russian coast wintering off Cape Farewell at the southern tip of Greenland. For some of this long trip they drift passively with currents, but for much of it they are actively swimming.

Razorbill
Alca torda
Smaller than either of the two Guillemots, the Razorbill has a breeding distribution confined to the coasts bordering the North Atlantic, and does not breed further north than the low arctic. It nests on the coasts of

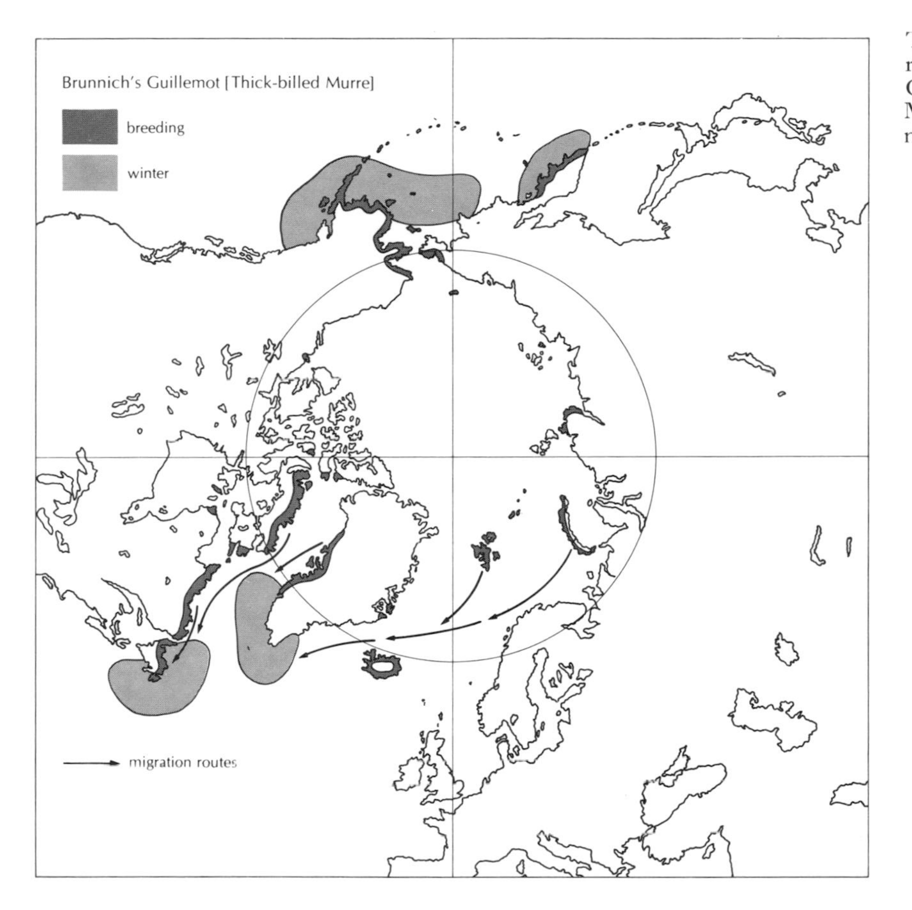

The breeding and winter ranges of Brunnich's Guillemot [Thick-billed Murre], and its principal migration routes.

Newfoundland and Labrador, western Greenland, Iceland, the northern and western coasts of the British Isles, on the Brittany coast, round much of Scandinavia including the Baltic, and it just reaches Russia on the Kola Peninsula. In winter it lives on the sea, not too far from land, on both sides of the North Atlantic as well as in the Mediterranean and off West Africa. The birds nest in crevices and at the back of deep ledges. They are found both high up on cliffs and among the boulder tumbles at the bottom. Their breeding habits are similar to the Guillemots, with an incubation period of about four and a half weeks for the single egg, and the young leaving for the sea at about three weeks old, when they have a waterproof plumage but cannot fly.

Black Guillemot [**Mandt's Guillemot**] including **Pigeon Guillemot**
Cepphus grylle
This is one of the very few bird species that can be found in the high arctic in the depths of winter. It seeks out leads in the pack ice, and areas round headlands and between islands where strong tides and currents keep some water open. In these it dives for its food of fish and bottom-living molluscs and crustaceans. As an adaptation to life in the extreme cold, the Black Guillemot has developed a winter plumage very different from its nearly all-black breeding dress. It changes to a largely grey and white plumage with only the back remaining black (even there then feathers there are edged with white) and with the whiteness concentrated under-

The breeding range of the Black Guillemot.

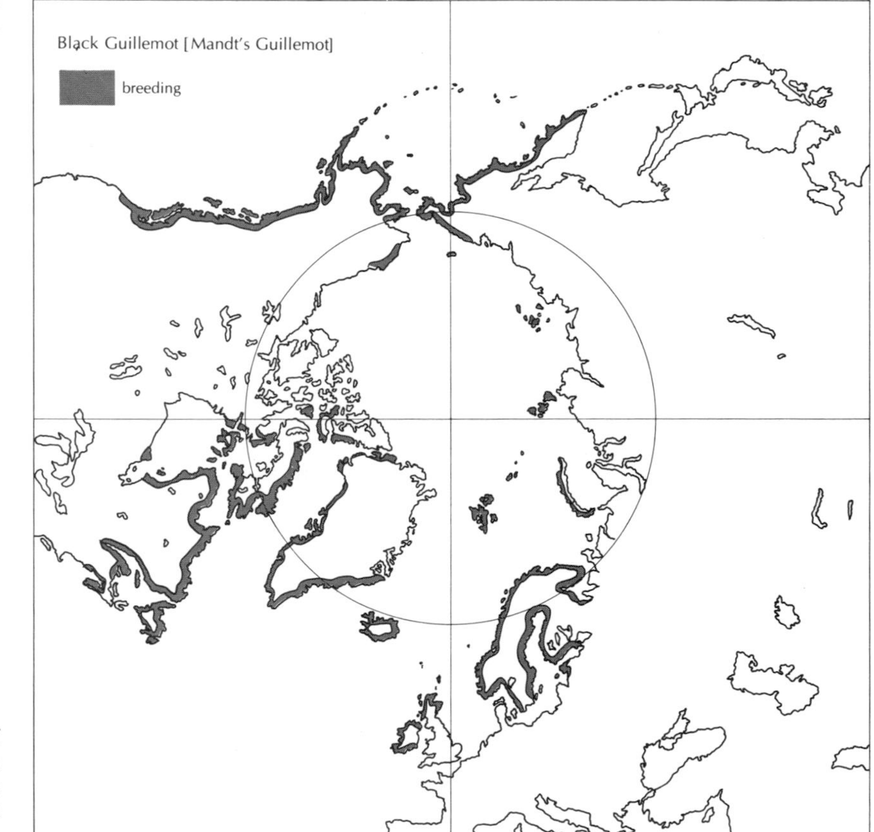

Opposite: A pair of Razorbills displaying out in the open. Their nest will be in a crevice or at the back of a deep ledge.

neath where it will be of most use. Almost all the auks increase the amount of white on their plumage in winter, but only the Black Guillemot and the Tufted Puffin, which are black underneath in their summer dress, change completely. However, even in summer this species has a white wing patch, which becomes larger in the birds breeding further north.

Black Guillemots nest in deep crevices and among boulders and rocky screes at the foot of cliffs, sometimes in loose colonies but never in the density of the other auks. The one or two eggs are incubated for about four weeks, and the chicks fledge in about a further five. Unlike other auks, the young are nearly full grown and can fly by the time they leave the nest. The juvenile plumage is very like the adult's winter dress because these youngsters will be spending their first winter in the arctic.

Auklets and Puffins

There are eight species in this group of which three, the Least and Parakeet Auklets and the Tufted Puffin, occur only in one area of the subarctic, just penetrating into the low arctic. The others are all found in the arctic, some exclusively so. They are closely related to the previous group of auks, but are all smaller, often markedly, and the smaller kinds are dependent on plankton for food rather than fish. They all nest in burrows and rock

crevices, lay only one egg, and rear their young to fledging before it leaves the nest. The large number of auks, auklets, and puffins breeding in northern lands is an indication of the richness of the seas there. Although there is considerable overlap in the foods taken by the different species of similar size, there are wider variations between the three main groups. The fish-eating Guillemots and Razorbill have a narrow bill, a rather hard tongue which helps to grasp their slippery prey, and a few very sharp denticles (bony protuberances) in their palate. The purely plankton-eating auklets have a much wider beak, a fleshy tongue, and a broad palate with numerous very small denticles, all of which helps them to take in a large amount of plankton and to filter the food particles efficiently from the water. In between these two groups are a few intermediates which feed partly on fish and partly on plankton. For example the Puffins eat plankton normally but feed fish to their young, presumably because they find this an easier food to carry in bulk back to the nest. Because the other plankton eaters are much smaller, they will not suffer so much from this problem.

Kittlitz's Murrelet
Brachyramphus brevirostris
Kittlitz's Murrelet is one of a group of three small auks or auklets that only occur on either side of the Barents Sea. They presumably survived the ice ages in the Beringia refugium and have not managed to spread any further. There is only one small auk outside this area, the Little Auk, which must have spent the ice ages somewhere in the North Atlantic from where it has been able to spread both east and west to a limited extent. Kittlitz's Murrelet now nests from Wrangel Island westwards along the Siberian coast to the Chukotski Peninsula, in western Alaska, and probably in small numbers along the north coast. It moves very little in winter, just to the southern part of the breeding range which stays ice free.

The birds breed on rocky coasts and islands, usually concealing their nests in rock tunnels beneath boulder slopes, but some are apparently more in the open. They have a particular preference for the vicinity of glaciers in Alaska where they choose adjacent rocky slopes, which can be up to five miles from the sea. Very little has been recorded about their breeding habits though the single egg is certainly incubated by both parents. Neither incubation nor fledging periods are known.

Little Auk [**Dovekie**]
Plautus alle
Quite a small bird, the Little Auk is about the same length as a Starling, though differently shaped, being stocky with virtually no neck

and only a short stubby tail. They breed in the high arctic between northwest Greenland and the Severnaya Zemlya islands off the coast of Siberia, and almost everywhere they occur in fabulous numbers. Their nesting sites are scree slopes which sometimes stretch for miles along the foot of mountain walls. One in Spitsbergen runs for about ten miles, forming a continuous colony, and sample counts have indicated the total number of birds there to be between three and five million. It has already been described how such vast numbers greatly enrich the vegetation below the colony, and here grows lusher, more dense vegetation on which geese as well as Reindeer graze, getting more food from small areas of enriched ground than they could from much larger areas of ordinary tundra. Little Auks lay just one egg, deep underneath the boulders making up the scree. The incubation period is twenty-four days, and the young fledge in four weeks. Their main predators are Glaucous Gulls and arctic foxes, but the predation level is extremely low relative to the enormous numbers of birds in the colony.

Little Auks [Dovekies] at their colony overlooking a Spitsbergen glacier. Some colonies contain several million birds.

Crested Auklet
Aethia cristatella
Least Auklet
Aethia pusilla
Parakeet Auklet
Cyclorrhynchus psittacula

It is convenient to deal with these three species together, partly because their ranges overlap to a considerable extent, and partly because all three have been studied on St Lawrence Island, Alaska, where they are found breeding in the same area. The studies have particularly highlighted how three apparently very similar small auks are in fact segregated in their breeding and food requirements. All three breed on the islands in the Bering Sea, and on the coasts of North America on the one hand and eastern Russia on the other. The Crested Auklet reaches furthest north and is really the only true arctic species of the three. It is found on the northern arctic coast of Alaska and probably also breeds on Wrangel Island off the north Siberian coast. The other two species have their mainland headquarters on the Chukotski Peninsula and on the southern and western coasts of Alaska, while the main colonies of all three are on the many islands and island groups stretching between the two continents.

The auklets can all be found breeding on and around large scree slopes on St Lawrence Island. Colonies of the Crested Auklet vary between 3,000 and 10,000 birds, while the Least Auklet occurs in greatest numbers with colonies from 5,000 to 200,000. These two species nest in among the boulders of the scree slopes but are topographically segregated by their preferences for different boulder sizes, the Crested Auklet liking large irregular ones, while the Least Auklet is

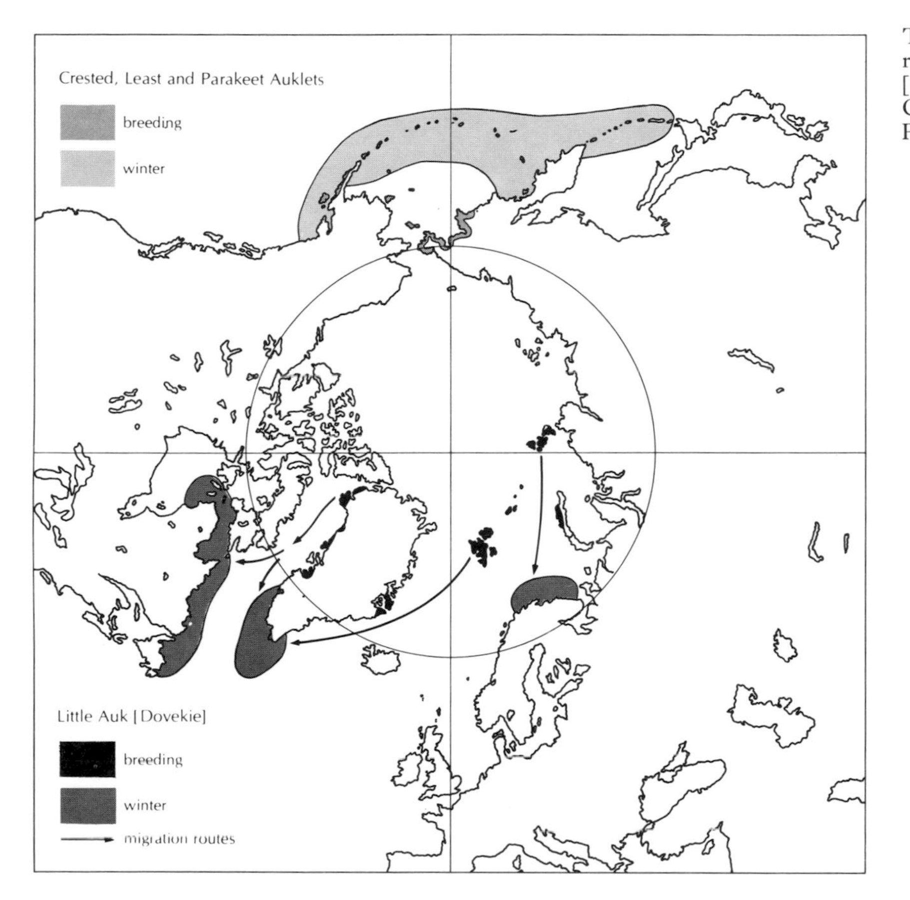

The breeding and winter ranges of the Little Auk [Dovekie] and the Crested, Least, and Parakeet Auklets.

much commoner among the smaller, more even scree. By contrast the Parakeet Auklet avoids the pure scree and nests in much smaller numbers (never forming large colonies) among the shattered pinnacles and rock ridges where the frost action is at its most vigorous. Thus three species of a similar type have managed to find separate niches for themselves in what might superficially appear a rather uniform habitat.

They all nest deep down among the rocks or boulders, as well hidden as possible. Great faithfulness is shown by individuals to the same site year after year, and in a season when the snow melts late from one particular area the birds will not move to a cleared place but stay by their traditional spot waiting for the thaw. Sometimes this produces a wide disparity in egg-laying dates, with some pairs able to lay two or three weeks before others. Conversely, in a late spring the thaw can be quick and nearly simultaneous over the entire locality, so that virtually all the birds are able to lay together.

It has already been described how geese and other species will resorb their egg follicles in a late spring, partly because they need the energy reserves the eggs represent and partly because delayed breeding is no use to them with their long cycle. Auklets on the other hand have access to abundant food in the sea, however snowcovered the land, and their breeding cycle is at most two or two and a half months, so they are able to delay their egg laying for some weeks without disastrous effect. Far from resorbing their eggs they are

In temperate latitudes Puffins excavate burrows in the soil. In the arctic where the ground is permanently frozen they make do with rock crevices.

sometimes unable to stop development, and in late spring some eggs get laid on the snow as close as possible to the site of the buried nest. In one bad year it was estimated that as much as 5% of the Least and Crested Auklet populations did this, with of course consequent egg mortality and loss of production. The habit has not been recorded for the Parakeet Auklet which lays rather later than the other two species, and so never has to wait quite as long as them.

Segregation between the three species as regards their plankton diet is achieved by differences in the bill size, which though small, are sufficient to prevent them enjoying the same food items. As its name implies, the Least Auklet is the smallest in size; indeed it is by far the smallest of all the auklets being only 5½ inches long (about the size of a House Sparrow) compared with the 8 inches of the Little Auk. Not unnaturally its prey consists of smaller items than the other two larger species will bother with. The Crested and Parakeet Auklets are about the same size, weighing rather more than half a pound each. However they appear to have inborn preferences for different animals in the sea plankton, backed up by small structural differences in their bills and differences in the depths to which they prefer to dive for their food.

All the birds lay just one egg, which is incubated by both parents in turn. The incubation and fledging periods of the Parakeet Auklet have been worked out at about five weeks each, but figures are still lacking for the other two. However, the Crested Auklet is likely to be similar, and the Least Auklet probably has shorter periods. The young of all three do not leave the nest until they are ready to fly. They then move out to sea with their parents, though the wintering areas are not far south from the breeding grounds, just into the continental shelf waters of the North Pacific.

Common Puffin
Fratercula arctica
Horned Puffin
Fratercula corniculata
Tufted Puffin
Lunda cirrhata

This is a second group of three species which are either close relatives or have such marked similarities as to make their treatment together more convenient than separately. Here, one species lives in the North Atlantic while the other two replace it in the North Pacific. The Common Puffin breeds in Labrador, Newfoundland, Greenland, Iceland, Britain, Norway, Bear Island, and Novaya Zemlya, and it winters at sea, mainly out as far as the continental shelf. The Horned and Tufted Puffins both breed on the Bering Sea island groups and on the mainland coasts on either side, but whereas the Tufted Puffin does not reach further north than the west coast of Alaska, the Horned Puffin is an arctic species, occurring on Wrangel Island and the Chukotski Peninsula. There is an area of overlap between them where they apparently coexist without competition but their true ecological relationships have not been studied.

Where there is sufficient soil available, all three puffins nest in burrows which they excavate themselves, sometimes up to ten feet long. The ground in a dense colony can become so undermined with burrows that walking over it there is a constant risk of putting a foot through into a hole. However, if there is permafrost near the surface or no soil they are content with natural crevices. The preferred nesting place is a steep, grassy slope with a fair depth of soil, such as occurs on the tops of small islands and sometimes halfway down precipitous cliffs. There is a direct relationship between the density of burrows and the degree of slope but on flatter ground, where the birds are sometimes forced to nest because of lack of other areas, the density falls away rapidly with increasing distance from the edge of the cliff. The reason is that the breeding success is much higher

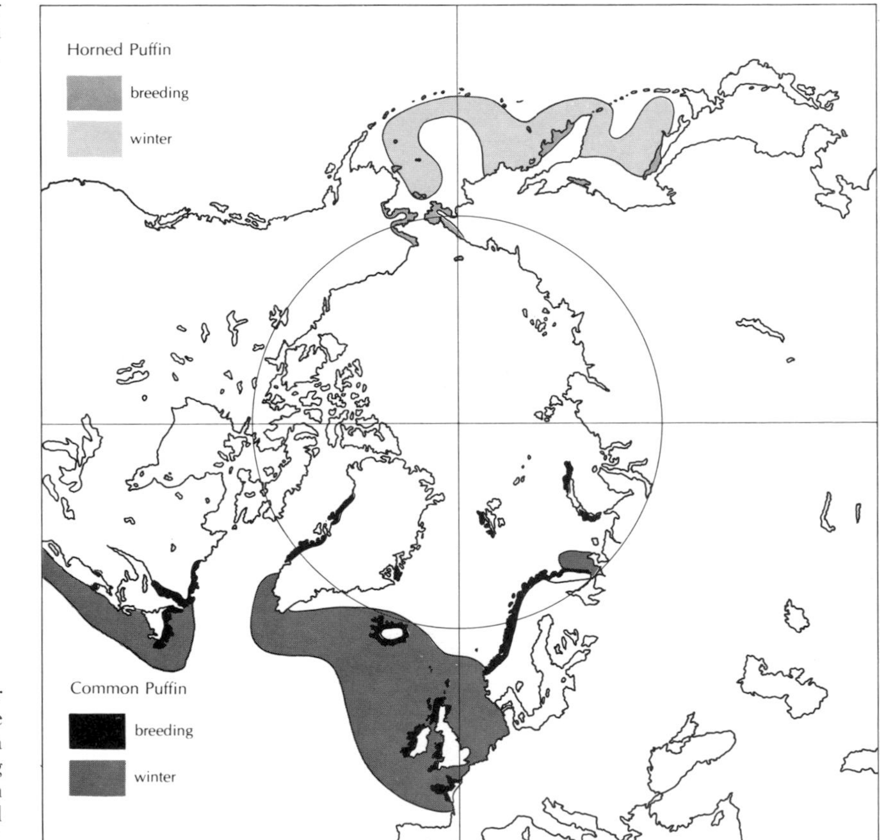

The breeding and winter ranges of the Common and Horned Puffins.

Opposite: The Fulmar has a remarkable defence mechanism against intruders, being capable of spitting an evil-smelling oil several feet with great accuracy.

in colonies on slopes than it is in colonies on flat ground. First the hatching success is greater because the loss of eggs is less. Puffins nesting on flat ground are much less settled than those on the far safer slopes, from where they can take off more easily and where predators find it harder to reach. As a result they suffer from more panic flights with every bird suddenly, and with little reason, taking to the air, and when this happens eggs are quite likely to get kicked out of the burrow. Secondly, in the week or two before fledging when the chicks begin to get hungry as the parental feeding tails off, they come to the mouth of the burrow, and it is easier for a gull to take one from flat land than from a steep slope.

All three species lay just one egg. The incubation period is about forty days, and the fledging period much the same. However the parents desert the chick before it has fledged, and leave it to find its own way out to sea. The pangs of hunger gradually force the young one first to come to the mouth of the burrow, and then to take the plunge and go to sea in search of its own food.

Fulmar

Fulmar
Fulmarius glacialis
The Fulmar is a cliff nesting bird belonging to the family of Petrels, most of whose other members are hole or burrow nesting birds.

It breeds in both the North Atlantic and North Pacific regions, mainly in the arctic but extending south as far as the British Isles, and in some of the island chains off the Russian Pacific coast. The principal haunts, however, are in Iceland, Greenland, Spitsbergen, Franz Josef Land, Novaya Zemlya, and some scattered places in the eastern arctic archipelagos. On the Pacific side it breeds on the Kurile island chain, on a number of islands in the Bering Sea, and on a solitary mainland site on the Chukotski Peninsula.

As an arctic bird it is of considerable interest because of its colour phases which vary from light to dark, the latter sometimes called 'blue'. There are numerous intermediate forms, but it is possible to classify most birds to either of the extremes, or to one or two intermediates. The occurrence of the different phases varies from colony to colony with a fairly regular north-south progression in both Pacific and Atlantic. However the two progressions run in opposite directions. In the Pacific the bird appears to obey the general rule of getting lighter the further north one goes, as an adaptation to help conserve heat: colonies in the Kurile chain (about 50°N) contain all dark birds, those in the Bering Sea north of 60°N have all light birds, while those in between have varying proportions of each. In the Atlantic, though, the southernmost colonies round the British Isles and in Iceland (between 50°N and 65°N) have all light birds, while those in the furthest north, in Spitsbergen and Franz Josef Land (77°N to 81°N), comprise all dark. This reversal in the Atlantic has produced various theories, but none that has also explained the Pacific arrangement.

In the last hundred years this bird has undergone a major spread round the North Atlantic, particularly in the British Isles where it was once confined to the St Kilda group of islands off the Outer Hebrides but where it now breeds almost all round the coasts. In the same period it spread right round Iceland, again from an original few sites of much greater age, and it also colonised the Faroes and much of Norway. The widely accepted reason for this enormous range increase is that the Fulmar, normally a plankton feeder, learnt to exploit the waste products of man's fishing industry, especially the guts and other scraps thrown overboard from the trawlers, and the similar rich food sources to be found in and around fishing ports. It is usual for the young birds in a population to found new colonies and to therefore spread the range. The breeding adults are attached to traditional sites but the young birds are free to wander and explore, and in favourable circumstances to spread, and the presence of an abundance of food would certainly allow more young Fulmars to survive the period between fledging and maturity, no less than eight or nine years, so that more potential breeders would be entering the population each year.

Fulmars lay only one egg, which the parents incubate in turn in spells of about four or five days for a total of about fifty-five

The Cormorant's nest is a large affair built of seaweed and other debris, and placed on flat rocks or broad cliff ledges.

days, and the chick does not fledge until it is about forty-six days old. This lengthy breeding cycle can be readily fitted into the short arctic summer because the birds nest on cliff edges which rarely get snow covered, and rely on the sea for their food, being prepared to fly a hundred miles or more to reach open water in the spring. It is for this reason that each parent incubates for such long periods. As already mentioned, the young bird does not breed until it is eight years old. Once it starts, however, it may go on for twenty or perhaps thirty years. Their true life expectancy is still not known despite long-term studies of ringed birds. One such study in the Orkneys began in 1950 and has shown that birds ringed in that year as breeding adults were still breeding twenty-five years later, so it is hardly surprising that this is such a successful species.

Cormorants

Cormorant
Phalacrocorax carbo
Pelagic Cormorant
Phalacrocorax pelagicus
Shag
Phalacrocorax lucidus
The three members of this genus are fish-eating birds, often taking flatfish from the sea bottom. They nest in colonies on broad cliff ledges or in hollows at the foot of cliffs, but Cormorants occasionally breed in trees in inland situations. All construct a bulky nest of twigs, seaweed, and various flotsam, the male generally collecting the material while the female does the building. They lay clutches of between three and five eggs which both sexes incubate for about thirty days. The young grow only slowly and do not leave

The Shag has an erectile crest and lacks the white face patch of the Cormorant. Their ranges overlap to a considerable extent.

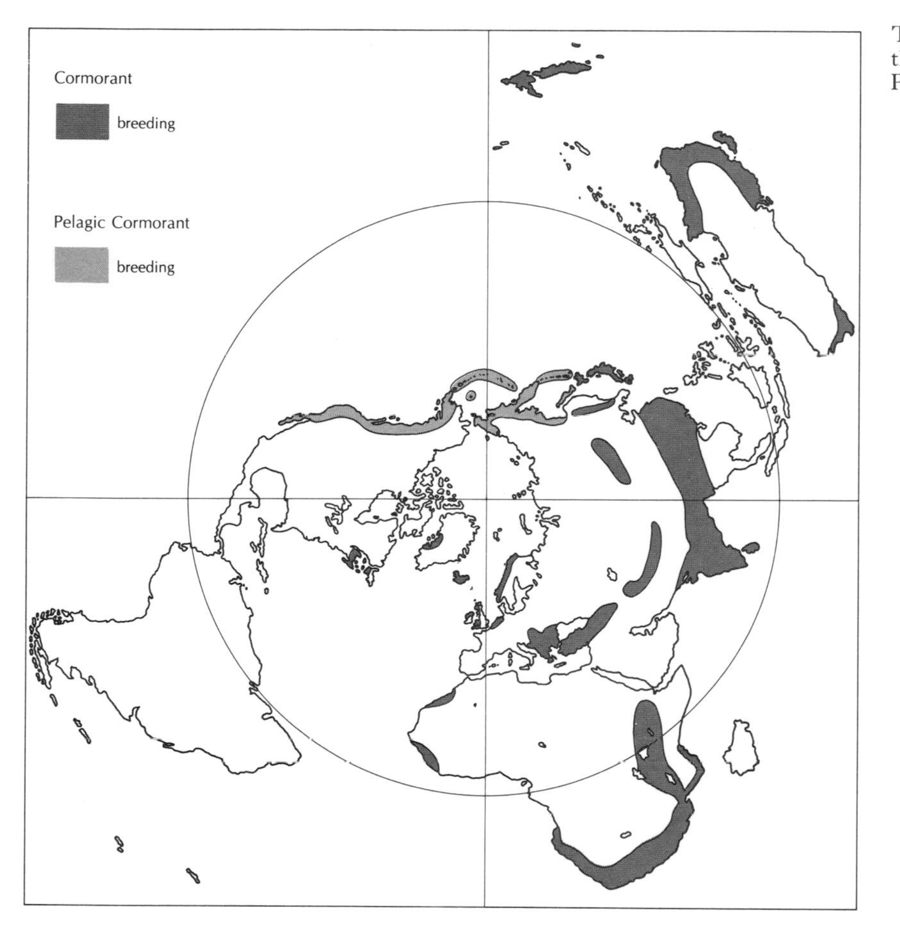

The breeding range of the Cormorant, and the Pelagic Cormorant.

the nest until they are fifty to fifty-five days old. Even then they may not be able to fly, and in any case remain dependent on their parents for food for a further period of a month or more.

All three birds only just reach the arctic, having the bulk of their breeding distribution well in the south. The Cormorant and Shag are confined to the North Atlantic where they breed in Iceland and on the northern coasts of Norway, and the Cormorant also breeds in western Greenland. The Pelagic Cormorant is confined to the Pacific and occurs on most of the island groups in the Bering Sea, plus the coats of Siberia and North America. Its furthest north is Wrangel Island, well into the arctic, and it also breeds on the Chukotski Peninsula. The species all winter either within their breeding range or, in the case of the arctic breeders, just to the south.

6 Land birds

The land birds consist of five widely separated orders that have no close relationship in the taxonomic sense, but all happen to live on the land throughout the year. They have no special affinity with water in that they neither swim on it nor dive in it, nor even normally wade in it. Some may get their food from the margins of wet places, but rarely exclusively so. The five orders are the raptors, owls, gamebirds, cranes, and passerines or perching birds. The first two groups are predators, some dependent to a large extent on the lemming and therefore subject to population and breeding fluctuations directly related to those of their prey. The latter three groups are plant and insect feeders, and they include among their number the most northerly breeding and wintering birds in the world.

Raptors

There are five arctic raptors or birds of prey, two of which extend well to the north, the other three being more southerly. They consist of one eagle, one buzzard, and three falcons. The eagle has the widest diet and is not therefore generally affected by food shortages, but the other species are all more reliant on one food type and their breeding success in any year will be increased or decreased depending upon the relative abundance of their prey. They are capable of producing more eggs in a good year and of rearing them all if the food supply is maintained, but if it slackens off then only some of the brood will survive. By having this inbuilt flexibility they are able to maximise production in a way that a species with a fixed clutch size is not able to do.

White-tailed Eagle

Haliaeetus albicilla

This magnificent bird has regrettably decreased in recent years but can still be found in southwest Greenland, Iceland, northern Scandinavia, and along the arctic coast of Russia; it also extends south through the boreal and temperate zones of Europe and Asia, breeding as far south as the Mediterranean and Caspian Seas. It is not a true arctic species, having clearly extended its range northwards, and is only able to survive in the northern areas because of its catholic diet. In winter, the birds move away from their Siberian breeding areas and come much further south, but they are resident in Greenland and Iceland, staying on open areas of coast.

White-tailed Eagles normally nest in trees, forming an enormous pile of branches and twigs which is added to each year, so that the nests may grow as large as six feet in diameter and four or five feet in height. The pair may have two or three nests on their territory and will use them in succession in different years. In the arctic regions they breed on cliff ledges on the coast or rarely on the ground on small islands. The usual clutch is two eggs, but there can be up to four,

The Fulmar population has increased and spread in the North Atlantic in the last 100 years, following its successful exploitation of waste discarded by fishing boats.

Tufted Puffins are confined to islands in the Bering Sea and the coasts of northeast Russia and Alaska. They shed their tufts in winter.

The unmistakable 'barn-door' shape of a soaring White-tailed Eagle. They have declined in many parts of their range, through persecution and poisoning.

which are laid at intervals of from two to four days, and incubation begins with the first egg. Both parents are thought to incubate (about thirty-five days for each egg), though the female does the larger share, and occasionally all of it. The young hatch at intervals of two to four days so the entire clutch will hatch about forty to forty-five days after the start of incubation. Throughout the fledging period the chicks usually remain different in size, but sometimes the youngest gradually catches up, especially in years when there is an abundant food supply. Unlike other birds of prey, however, it is extremely rare for the youngest and smallest to starve and die in years of poor food supply. This is partly because the clutch never exceeds four, and the parents seem able to keep this number supplied with food in all conditions, and partly because they have such a wide spectrum of food items on which they prey. Fish form an important part of their diet, the birds plunging feet first into the water to grasp them. Many species of animals and birds are also taken, including young deer, hares, voles, geese, ducks, gulls and ptarmigans.

When the young are small the female stays with them, brooding and guarding them, while the male hunts and brings food to the nest for all the family. Later both parents hunt, tearing up the prey and feeding small pieces to the chicks, but they soon learn to tackle the feeding themselves. They leave the nest when about eight weeks old, in spite of scarcely being able to fly at this stage, and hop and flap about. Flying comes after another fortnight, though they are still dependent on their parents for food for a further five or six weeks.

Rough-legged Buzzard [Rough-legged Hawk]

Buteo lagopus

The Rough-legged Buzzard has been divided into four races, but together they complete a circumpolar distribution that has just one gap, in Greenland and Iceland. It does not reach the far northern islands of Eurasia, nor the most northern islands of the Canadian arctic. The bird is a migrant, wintering to the south of its breeding range, mainly on open fields and marshes in temperate lands.

Throughout its range, the Rough-legged Buzzard feeds on small mammals, especially lemmings and voles, and its breeding cycle is closely linked to the abundance cycles of these animals. The nest is constructed of twigs, grasses, and other vegetations, and is usually placed on a cliff ledge or rocky outcrop. In some areas a low mound above the general tundra level will suffice. The clutch size varies from one to as many as seven, but in an average lemming year it will be three or four. Incubation starts with the first egg laid and takes about twenty-eight days for each one. Both sexes participate, though the female probably does the larger share, and the young

The breeding and winter ranges of the Rough-legged Buzzard [Rough-legged Hawk].

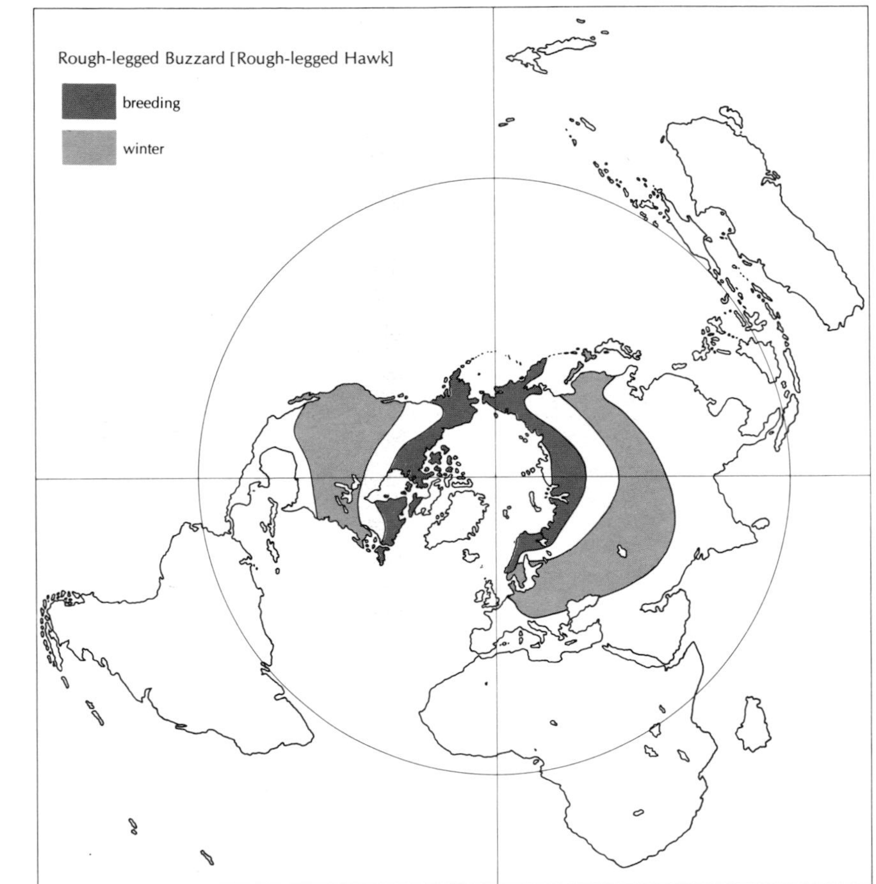

hatch at intervals of about two days corresponding with the egg-laying interval. The female stays at the nest for the first week or two, the food being brought by the male, but after about two weeks she ceases to feed each young in turn and merely places the food in the nest for them and starts hunting herself. It is at this point that mortality may occur among the chicks. In good lemming years all will be successfully reared but if after an initial favourable period the number of lemmings decreases sharply, as can happen because of bad weather or disease, the parents will be unable to bring in enough food to rear all of their brood. In this situation the youngest ones will starve and may even be eaten by their older siblings, for it is better that one or two healthy chicks be reared than that all the brood should become weak and eventually die. The fledging period is about forty-one days, though they may move short distances away from the nest before this.

Gyr Falcon

Falco rusticolus

The most northerly living of the three falcons, this bird breeds in the high arctic right round the pole. It is absent from the arctic islands north of Eurasia but occurs on virtually all those of the Canadian arctic. It is a partial migrant, moving from inland to the coast, and southwards to an irregular extent. For centuries this has been the prized bird of falconers, sought by kings and emperors as befitting to their station. The recent revival in falconry has put pressure on some of the birds of prey at a time when other pressures such as human disturbance and pesticides have already reduced their numbers.

The Gyr Falcon has a number of colour phases, sometimes differentiated as subspecies, but now generally regarded as mere variations. There are three—white, grey, and brown—with complete intergradation between them. The white form is found almost throughout the breeding range, though not in Iceland or Scandinavia. It is commonest in the north and may be regarded as an adaptation to the cold. In Greenland virtually all the birds are white while in North America they vary from between 4% and 50% as one moves from south to north. The brown type is commonest in Labrador and on the Kamchatka Peninsula on the Pacific coast of Russia.

The birds retain the same territories year after year, sometimes returning to the same nest site, but more often alternating between two or three different ones. The usual site is an inaccessible cliff ledge, protected by an overhang. Here the parents construct a rudimentary nest of twigs which over several successive years becomes built up into a larger structure, often incorporating the remains of their prey. In areas where there are no sea cliffs the birds will move inland to find river gorges or mountainous areas, or else trees at the edges of forested regions, where they often take over an old nest of a raven or buzzard.

The Gyr Falcon's breeding range.

The Gyr Falcon breeds far into the high arctic. The further one moves north so the proportion of white birds increases.

The normal clutch is three or four eggs, but between two and seven have been recorded. The smaller and larger numbers relate to the abundance of the food supply, just as in the Rough-legged Buzzard, but here it is the Willow Ptarmigan that is important. In some areas the lemming and arctic hare are also major food sources. However, many pairs of Gyr Falcons nest on the coast near large seabird colonies, from which they take all their prey. These falcons therefore have a constant food supply and are able to lay consistent clutches of normal size and to rear the maximum number of young. Among the other broods losses from starvation and perhaps cannibalism will occur in time of shortage. The eggs are incubated by the female only, starting with the first egg. She sits for about twenty-eight days before the first one hatches, and then stays with the brood fairly continuously for the first fortnight. Fledging takes place when the young are about seven weeks old, and independence comes after another month.

Peregrine Falcon [Duck Hawk]
Falco peregrinus
The Peregrine is a close relative of the Gyr Falcon and their breeding ranges overlap to a considerable extent, though the Peregrine also breeds almost throughout the world and in a very wide variety of habitats. As many as twenty-two subspecies have been named, most restricted to certain areas of the globe, but only two need be mentioned here. The race breeding in the Eurasian arctic, including some of the islands to the north of the Russian coast, is *calidris*, and it is rather larger and paler than the nominate *peregrinus*, which breeds in Britain and northwest Europe. The race *anatum* is also larger than the nominate form but somewhat darker, and it lives in arctic Canada and parts of west Greenland. Clearly there must be some separation that allows two such similar birds of prey as the Gyr and Peregrine Falcons to live in the same areas, and the answer is their food. Although both are predators, striking down birds from the air with their powerful talons and capable of picking quite large birds or animals from the ground, the Gyr takes mainly Willow Ptarmigan plus lemmings and hares, or in some cases seabirds, while the Peregrine takes a much wider variety of birds mostly rather smaller than the Ptarmigan. In a few cases however the diet is identical, for example those pairs which nest near seabird colonies, but here the teeming numbers of auks, kittiwakes, and gulls can supply the needs of both falcons without competition.

The breeding details of the Peregrine are essentially similar to those of the Gyr, but the nest site is a little more varied as it can be on a low mound, and is therefore accessible in a way that a Gyr's nest hardly ever is. While the chicks are small their main food is the downy young of other birds, chiefly waders and passerines, but as they grow so the food items get larger and include adult waders, ducks, and gulls. The extraordinary relationship between the Red-breasted Goose and the Peregrine in the Russian arctic has already been mentioned. It appears from this and other studies that the male Peregrine, which does most of the hunting, will not normally take birds from within about 200 yards of its nest. Quite why this should be is difficult to explain but the Red-breasted Geese certainly take advantage of it.

Merlin [Pigeon Hawk]
Falco columbarius
The only areas where this widespread boreal and temperate small falcon reaches the arctic are in Iceland and in restricted parts of the coasts of Russia and Canada. There are several named subspecies but the variations have little relation to arctic adaptations. Merlins nest on the ground, selecting a sheltered hollow in open moorland, sand-dunes, or marshes, but occasionally they breed on a cliff ledge or in a tree nest of some other species. The female incubates the clutch of five eggs, sometimes six, and is fed by the male during the thirty day period. She continues to brood the chicks while they are small but then joins the male in hunting. The young fledge in about four weeks and stay with their parents for a few more before becoming independent. The principal food

The Peregrine Falcon [Duck Hawk] takes smaller prey than the Gyr Falcon, thus avoiding competition in areas where they occur together.

of Merlins is small birds, generally taken on the wing. In the north the fledging period coincides with the time when most young passerines are leaving the nest and are thus easiest to catch. The female bird is about 10% larger than the male and consequently she takes rather larger items of prey, up to the size of a small duck or young Ptarmigan.

The smallest raptor breeding in the arctic is the Merlin [Pigeon Hawk]. It usually nests on the ground in thick vegetation.

Owls

There are two species of owl in the arctic: the Snowy Owl has a circumpolar distribution in both high and low arctic, while the Short-eared Owl is mainly a subarctic bird with some extensions in to the arctic proper. Both have similar breeding habits, particularly in relation to their main prey of lemmings and voles. The variation in clutch size is far greater than in the raptors, showing an even greater capacity to take advantage of peak lemming years.

Snowy Owl
Nyctea scandiaca
The only parts of the high arctic where this species is absent are some of the island groups such as Spitsbergen, Franz Josef Land, the New Siberian group, and so on, where lemmings and voles are either completely absent or very rare. Everywhere else it breeds right to the very far north of, for example, arctic Canada and Greenland. In winter it stays within the breeding range, though those in the extreme north move south to some extent, at least for the period of total darkness from about mid-November to February. The completely white plumage serves the double purpose of providing maximum heat insulation coupled with excellent camouflage to prevent detection by its prey. It also has less obvious physical modifications to combat the cold by having completely feathered feet and legs, and very thick pads on the undersides of its feet.

The nesting density and breeding capacity of the Snowy Owl is almost wholly dependent on the availability of lemmings. In poor years the Snowy Owls will either not breed at all or lay very small clutches of three or four eggs only. In peak lemming years, however, clutches of eight or ten are quite usual, and up to no less than fifteen have been recorded at exceptional times. The usual nesting habitat is lowland tundra where lemmings abound, but an essential requirement is a prominent or rocky outcrop for the nest and for the male's look-out point. The birds make little or no attempt to conceal themselves, and a male on watch is visible for miles across the tundra; indeed in a peak year white blobs can be seen in all directions and on almost every knoll. Some males vigorously defend the nest against an approaching human, even hitting him about the head—a very painful experience. Incubation is carried out by the female and lasts about thirty-two to thirty-five days, but as it starts with the first egg a large clutch can take a fortnight or more to hatch. To begin with the male supplies food for the chicks, with both parents hunting later. The young leave the nest at about four weeks old but do not fly for a further four weeks. Although occasional males are bigamous, supporting two females on nests within the same territory, this is probably only possible in good lemming years.

Some work in arctic Canada has demonstrated how many lemmings a family of Snowy Owls will eat, and what effect they have on the total population. This was done by observations at the nest and by finding out the average density of the rodents within the hunting areas. It was found that in the thirteen week period from hatching to independence, a pair of adults and their brood of nine ate about 3,300 lemmings, made up of about 1,500 for the young and 800 each for the adults. The numbers were approximate and varied with the average size of the lemmings, but the daily intake only a fraction of the number available per day—less than 0.2%— and the lemmings were of course breeding all the time, replacing at

The annual breeding success of the Snowy Owl is strictly controlled by its food supply. In poor lemming years few young will be reared.

least some of those taken. One other interesting point that has come to light in recent studies is that in periods of continuous daylight the birds still concentrate their hunting in the 'night', being more active in the period from 2100 to 0800 hours than in the rest of the day.

Short-eared Owl

Asio flammeus

The Short-eared Owl does not breed on any of the arctic islands, whether in Canada, Russia, Iceland, or Greenland. However it breeds right up to the arctic ocean in both Russia and Canada, as well as to the south into the temperate zone. Its breeding habits are similar to those of the Snowy Owl. In plague years of voles, its staple diet, it can lay clutches of up to fourteen eggs, but if the food supply is not maintained several chicks may starve or even be eaten by the older ones of the brood. The female incubates for about four weeks, and the young leave the nest at from twelve to seventeen days old. They cannot fly at this age but lurk in thick cover near the nest, attracting their parents when they return with food by means of wheezing and hissing calls. It takes them another fortnight before they can fly. In winter, the birds are frequent on the coast and on open areas of rough grass inland, hunting their prey even in broad daylight.

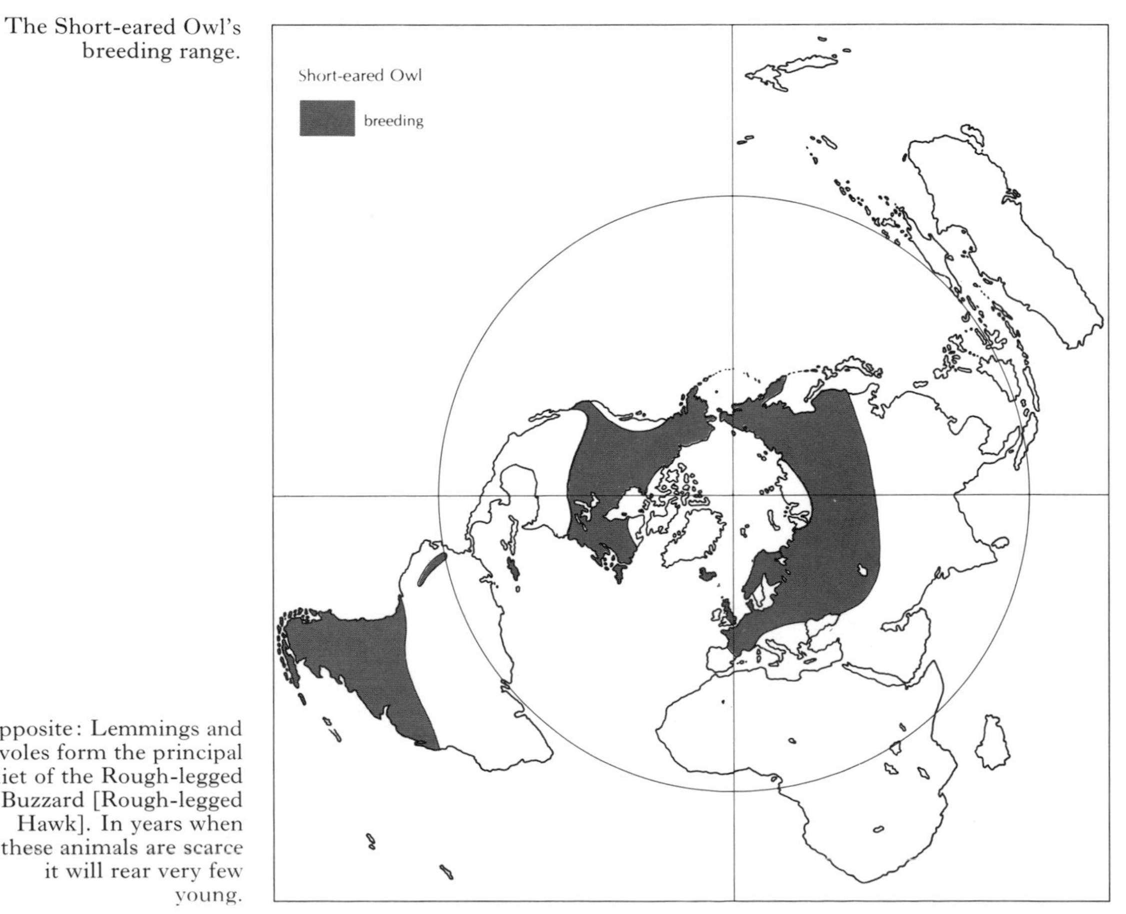

The Short-eared Owl's breeding range.

Opposite: Lemmings and voles form the principal diet of the Rough-legged Buzzard [Rough-legged Hawk]. In years when these animals are scarce it will rear very few young.

Opposite: The Savannah Sparrow lives in open habitat throughout North America, not just in the arctic. It is a weak flier and often just runs away if disturbed.

The commonest small bird of the arctic is the Snow Bunting. The male in his handsome black and white breeding dress has a distinctive moth-like display in flight.

Opposite: In North America the Shore Lark [Horned Lark] occurs widely in the Prairies and other open habitat, but in Eurasia it is confined to the arctic or to mountain ranges.

A Short-eared Owl with its main prey, a vole. In winter this species is found on the coast often hunting in broad daylight.

Gamebirds and Cranes

In this group are two true gamebirds, the Willow Ptarmigan and the Rock Ptarmigan. Both are circumpolar arctic species, though the latter goes very much further north than the former, which also occurs in wooded areas to the south of the tundra. Placed with them is the sole arctic crane, the Sandhill—this long-legged land bird seems to fit better here than anywhere else.

Willow Ptarmigan [**Willow Grouse**]
including **Red Grouse**
Lagopus lagopus
Essentially this is the low arctic Ptarmigan, with a complete low arctic breeding range except for Iceland. However there is extensive overlap with the Rock Ptarmigan, and separation is by the use of differing habitats: where the latter prefers the higher and drier uplands, often poorly vegetated, the Willow Ptarmigan seeks the wetter, well-vegetated tundra or the scrub zone to the north of the tree line. There are several races of this bird, including the Red Grouse of Britain and the Willow Grouse of Scandinavia, but the name Willow Ptarmigan seems to fit better bearing in mind its close relationship with the Rock Ptarmigan. The more northerly races moult into a white winter plumage, which benefits them both by its camouflage properties and by its insulation. The legs and feet are feathered.

Willow Ptarmigan are partial migrants, moving no further south than they have to. They arrive back on the breeding grounds in flocks and go straight to snow-free areas of tundra. Here the males immediately begin displaying towards the females, with their delightful and rather surprising courtship flight in which they fly low over the ground and then suddenly shoot up to a height of about 50 to 60 feet before descending rapidly to the ground, all the while uttering a loud cackling call. The females take remarkably little notice of them, spending almost all their time feeding to lay down the necessary energy for the production of the eggs. Gradually however the pairs sort themselves out of the flock and spread out to take up territories which are vigorously defended against other males and which average seven to twenty acres in extent.

The birds nest on the ground, forming a shallow depression lined with leaves and grasses. Their clutch varies considerably from a normal six to eleven to extremes of four to seventeen. Incubation, which takes about twenty-two days, is carried out solely by the female, whose summer plumage is beautifully cryptic. As soon as they hatch, the chicks are active and, in common with other members of their gamebird family, grow rather little in size during the first two weeks of life, instead devoting a large proportion of their energy into growing wing feathers. Thus at no more than twelve to thirteen days old they can fly, taking to the air on short, stubby, but perfectly effective wings. Once they have attained this ability to escape from possible

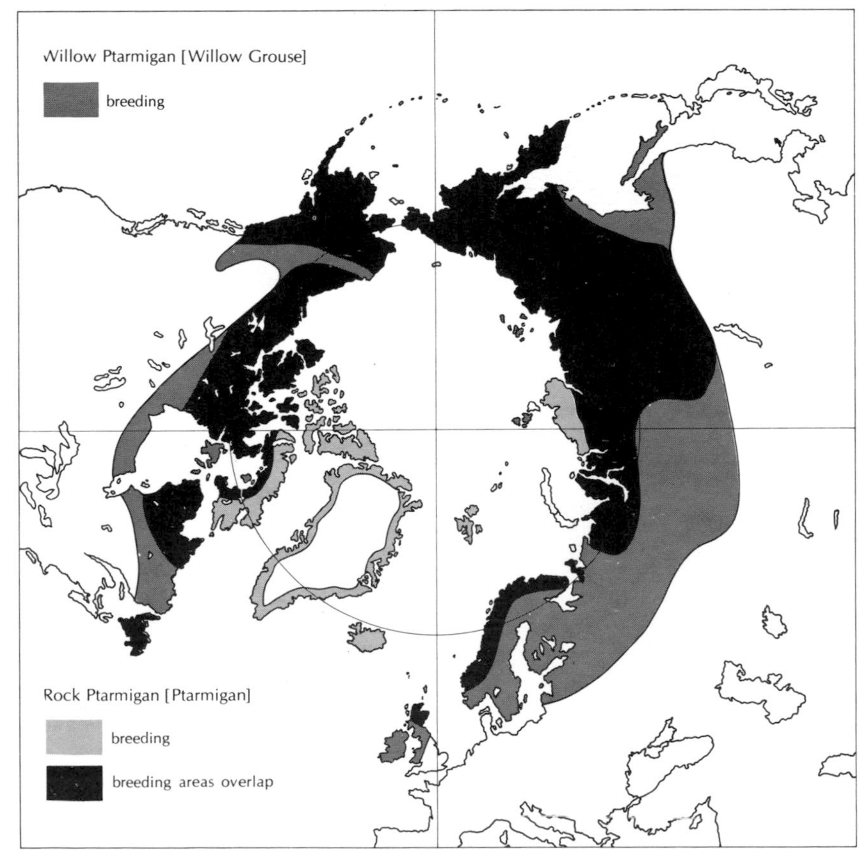

The overlapping breeding ranges of the Willow Ptarmigan [Willow Grouse] and the Rock Ptarmigan [Ptarmigan].

The near-perfect camouflage of the female Willow Ptarmigan [Willow Grouse] blends with its surroundings. She has to incubate for about three weeks.

predators, they get on with the business of growing to adult size, achieved in a further six weeks. In contrast, birds living to the south of the arctic grow much more slowly, not reaching full size until they are three months old, but then they lack the pressures of a very short summer.

Rock Ptarmigan [Ptarmigan]
Lagopus mutus
There are even more races of the Rock Ptarmigan than there are of the Willow Ptarmigan, a sure sign of a non-migratory species in which localised sedentary populations slowly evolve into separable races. It has a completely circumpolar distribution in the high arctic, occurring on most of the island groups in the far north as well as the mainlands, and it also extends south, living on mountain ranges of the Alps, Pyrenees, and the Altai.

The Rock Ptarmigan is one of the very few species to winter in the arctic, moving south only for the period of complete darkness, though even then some may stay in remote areas. To help survive the cold, and secondarily for perfect camouflage, the birds go completely white in winter. The pattern of moult has been studied closely and has been shown to be directly related to average temperatures. It begins when the average temperature sinks to about 4.5°C (40°F) and is completed by the time the temperature has dropped to between −2°C and −4°C (25°F and 28°F). This can be translated into approximate dates. For example in the high arctic part of Greenland the first temperature is reached in mid-August, and at about this time the Ptarmigan begin to go white. Their moult is completed by the second half of September, when the temperatures are already averaging below freezing. In low arctic

Greenland the corresponding dates are late August and late October. Further south again, for example in the north of Scotland and in the Pyrenees, the moult does not start until late September at the earliest and in the majority of birds it is never completed, for the average temperatures just do not go low enough to stimulate the birds into completion. They are at their whitest in January, the coldest month of the winter, but soon afterwards the moult back again begins. The strongest reason for believing that the winter moult is linked with temperature and not just with snow is that all the birds change into white plumage before the first snows appear.

Like Willow Ptarmigan, these birds rarely seem to stop moulting—the males have three or possibly four plumages in a year. They moult from their white winter dress into a spring plumage, and then with some small changes into their summer plumage. As breeding finishes, this is modified into an autumn dress, but before completion they are beginning to moult into the white of winter once more. Their wings are shed twice a year, in spring and in autumn. The full sequence of moults in both Ptarmigans is extremely complex and varies, as explained above, with latitude. The breeding habits too are similar to those of the Willow Ptarmigan, except that they never achieve such high nesting densities, being well scattered across the more barren tundra where they live. The usual clutch is eight to ten eggs, which the female incubates for about twenty-five days. The young can fly at about ten days old, but are not fully grown for a further five to six weeks. Ptarmigan of both species are vegetarians, taking plant buds, stems, and berries. The Rock Ptarmigan possesses a crop or food storage organ which can be filled during the short winter day and then the contents digested during the night. This greatly increases their total energy intake and is vital in enabling them to winter so far north.

The primary reason for the white winter plumage of the Rock Ptarmigan is insulation against the cold. It is secondarily good camouflage.

Sandhill Crane

Grus canadensis

The Sandhill Crane is confined to the North American continent except for a relatively small area of northeast Siberia along the north coast from the Kolyma river to the Bering Sea, and probably on Wrangel Island. It extends for some distance inland as well. In North America it occurs widely through northern Alaska and across the centre of arctic Canada, breeding on the more southerly of the islands, and extends south on the mainland into the boreal zone west of Hudson Bay. Another race, the Greater Sandhill Crane, breeds well to the south into the United States.

These long-legged, long-necked birds, with bodies the size of small geese, fly like geese in V formations or in long lines. They arrive on the breeding grounds—arctic tundra, river flats, and the drier areas of marshes—in late May or early June. The pairs quickly separate off to the accompaniment of their courtship dance, in which the two birds face each other with widespread drooping wings and shuffle and leap into the air. The nest is a mere scrape on a dry patch of ground, sometimes with only a handful of grass or twigs as a lining but often a substantial structure of available vegetation, especially in wetter areas. Just two eggs are laid and these are incubated in turn by both parents for a total of about thirty days. They are extremely secretive during the nesting period but by August, when the young have fledged, they gather in large flocks. Their migration takes them south through the prairie provinces and the United States to winter in southern California, Texas, and Mexico. In certain farming areas these autumn flocks, some containing 10,000 or more birds, do substantial damage to unharvested grain, stripping the heads and trampling the stems underfoot. Shooting has a limited effect and in any case has been severely restricted until recently because of the danger of killing one of the very similar, but very rare, Greater Sandhill Cranes, but scaring with bangers and the purchase by conservation agencies of fields left unharvested for the birds (which then do less damage elsewhere) have helped in many places.

Larks, Pipits, and Wagtails

These small birds, members of the passerine or perching bird family, have no problem fitting their breeding cycle into the short summer: the sum of nest building, laying, incubation, and fledging periods rarely totals more than six weeks. However being small has disadvantages as well as advantages, among which are the relatively greater energy intake needed to keep warm compared with larger birds, the long migrations needed to reach suitable wintering areas, and the greatly lessened ability to sit through spells of bad weather without feeding. The small birds that do live in the arctic are in general restricted to those which can live on plant food, even though insects will be taken in preference when available, and indeed several manage to be purely insectivorous in more hospitable parts of their range. Most of the species have to nest on the ground, but some use bushes in the low arctic, while a few conceal their nests in crevices in rocks. All are near the bottom end of a food chain, being liable to be eaten by various predators including raptors, owls, skuas, large gulls, and arctic foxes.

Small birds in temperate lands are able to have two or even three broods in a summer to overcome the high natural mortality they suffer, especially of eggs and young. Arctic passerines apparently never try to fit two broods into a summer but make up for this to some extent by laying slightly more eggs on average than do their relatives further south, with the longer hours of daylight allowing the parents to gather more food for

A pair of Sandhill Cranes stepping through a marsh. They are normally vegetarian but will sometimes take eggs and chicks of other birds.

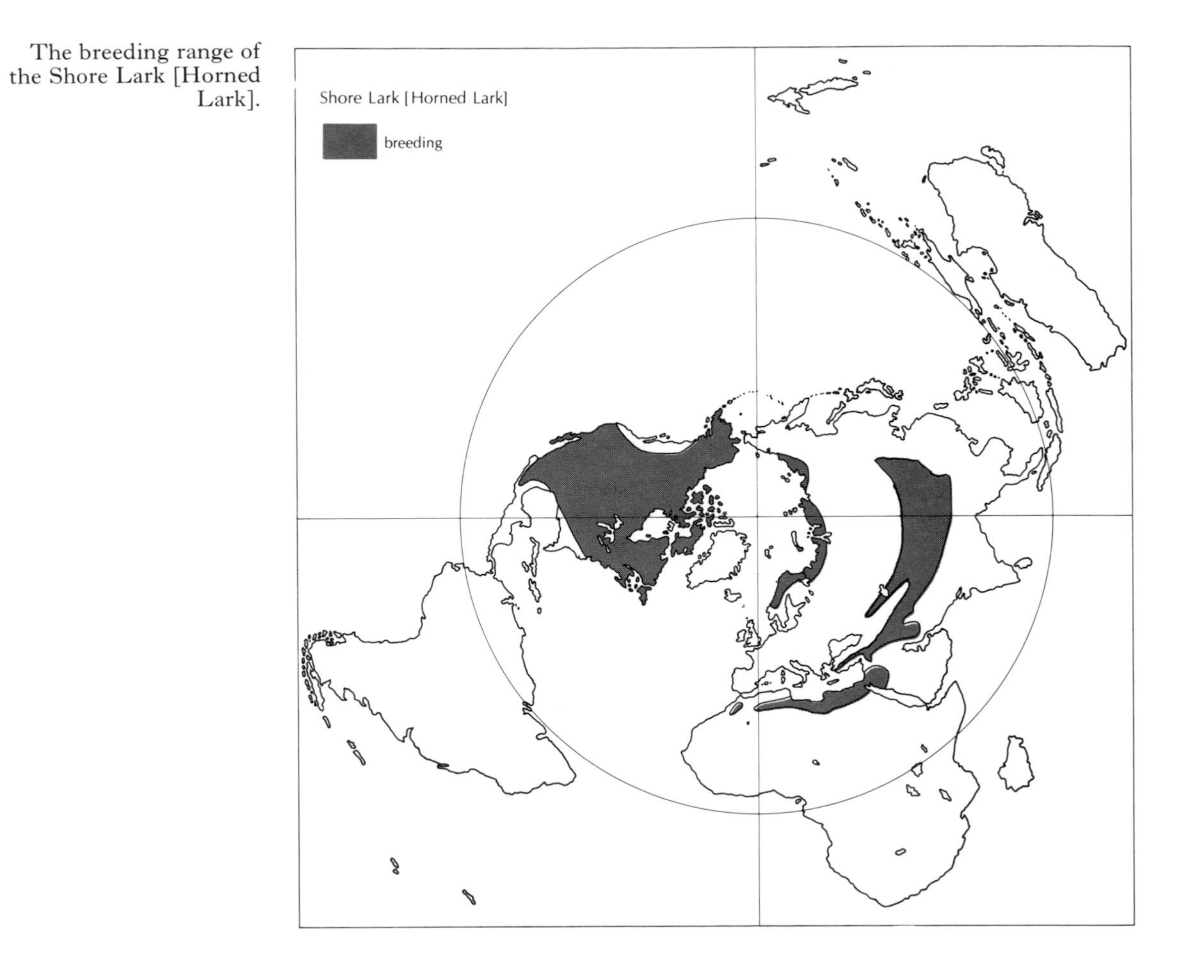

The breeding range of the Shore Lark [Horned Lark].

the extra mouths. Despite the predators, they seem capable of a higher rearing rate with the young, for one advantage of having only one family is that the parents do not desert the chicks soon after they have fledged in order to get on with a second brood, but continue to look after them. In this group are one lark, four pipits (of which two are primarily arctic), and three wagtails. None are exclusively arctic, but six of them have major parts of their range within its boundaries. All are insect feeders away from the arctic but all eat large amounts of plant food, especially seeds, small leaves, and buds when in the north.

Shore Lark [Horned Lark]
Eremophila alpestris
Taxonomists have named over a dozen races of the Shore Lark including three in the arctic. It has an enormous range, embracing almost the entire polar region with the exception of Greenland, Iceland, Spitsbergen, and a small area on either side of the Bering Sea. It also breeds virtually throughout the continent of North America extending right down almost to the Isthmus, and in addition there are separate populations in virtually all the mountain ranges of North Africa, the Middle East, through southern Russia to western China. In the far north this is a bird of barren

tundra; in the mountains it inhabits the bare stony lands above the treeline, spilling over into the dry upland steppes of central Asia. In North America away from the arctic it is found in habitat similar to that of the European Skylark in Europe, namely the short-grass areas of the prairies, ploughed or poorly vegetated land, and open expanses such as airfields. It is also found in American deserts (where there are no larks similar to the Eurasian Crested and Thekla Larks), and it must be assumed that it colonised these areas by spreading south from the tundra down mountain chains such as the Rockies.

The birds nest in dry areas of tundra, often gravel or stone flats completely free of vegetation. They also nest in heath vegetation and on upland grassy areas, but again only when they are dry. The pairs adopt nesting territories which the male demarcates by song, including a delightful, short song flight, and in one area of Canada these averaged between 15 and 25 acres. The nest itself is built of grass and other vegetation lined with finer material. It is placed on the ground, usually in a slight hollow, and often with a southerly aspect for maximum benefit from the sun. The clutch is usually between four and seven eggs, and these are incubated by the female for about twelve to fourteen days. In common with most ground-nesting small birds in the arctic, the young leave the nest before they can fly, at about eight or nine days old. They disperse in all directions and the adults locate them by their calling when returning with food. This abandoning of the nest increases their survival chances in the period when the brood is getting quite large and noisy, and therefore conspicuous to a predator—a solitary, crouching chick, silent except when its parents are near, is much harder to spot. In a further five days they are able to fly.

Shore Larks in their southerly desert and mountain haunts are sedentary, a fact that has helped contribute to the considerable subspeciation. Some arctic living birds migrate south to winter on the coasts of northwest Europe and in poorly cultivated fields inland. Nowhere are numbers great, and single birds or very small flocks are most usual. In North America, however, it is a relatively plentiful species and flocks of many hundreds occur in some areas. Here the arctic birds move south to join the resident birds living in the rest of the continent.

Red-throated Pipit
Anthus cervinus
The Red-throated Pipit is a Eurasian species with a distribution running without a break from northern Scandinavia across northern Russia to the Chukotski Peninsula. Its sole North American breeding record came in 1931 when a female and a clutch of four eggs were collected from the Seward Peninsula in Alaska. Since then a small number of individuals have been recorded there but no further breeding. It winters in central Africa, south of the Sahara, and in southeast Asia including southern China and Indonesia. The principal habitat of these birds is wet boggy tundra, preferably in willow and birch scrub of the low arctic, but in much lower vegetation in some areas of the high arctic. Their nest is a typical pipit cup made of interwoven grasses, lined with finer grass and hair, and is placed on the ground, usually half concealed in the side of a grass tussock or under a bush. Five or six eggs are laid which are incubated by the female for about twelve days, and the young are fed by both parents and fledge in eleven to thirteen days.

Water Pipit including **Rock Pipit**
Anthus spinoletta
There are two main groups of races in this species: the *spinoletta* group or Water Pipits, which are birds of tundra, mountain pastures, and open land, and the *petrosus* group or Rock Pipits, which live on rocky coasts and islands. The former occur widely in arctic North America, western Greenland, and in northeastern Siberia, while the latter reach

the arctic only round the Kola Peninsula and in the White Sea, and both groups occur extensively to the south of the arctic. Rock Pipits are virtually sedentary throughout their range, but Water Pipits move out of the arctic for the winter to the nearest ice-free land.

The Water Pipit is the more arctic of the two, and its breeding habits have been studied in Canada, for in parts of southern Baffin Island it is the third commonest among the small birds (after the Lapland and Snow Buntings). As might be guessed from its name, it prefers the wetter areas of tundra, but is also found on drier heathy places. The nest is a simple cup of grass, lined with finer grasses and always well concealed under overhanging vegetation, and the clutch of four to six eggs is incubated by the female. The birds show a liking for a warm sheltered slope for their nest site: on Bylot Island at the extreme north of Baffin Island the only nests found of this species were on the south-facing slopes of steep ravines, where they would capture the maximum amount of warmth from the sun. This area is close to its northerly limit and the bird was not nearly as common as further south.

In 1953 the Water Pipits breeding on Baffin Island suffered very severe losses from a spell of wet, cold, and foggy weather. It only lasted two days but was sufficient to kill large

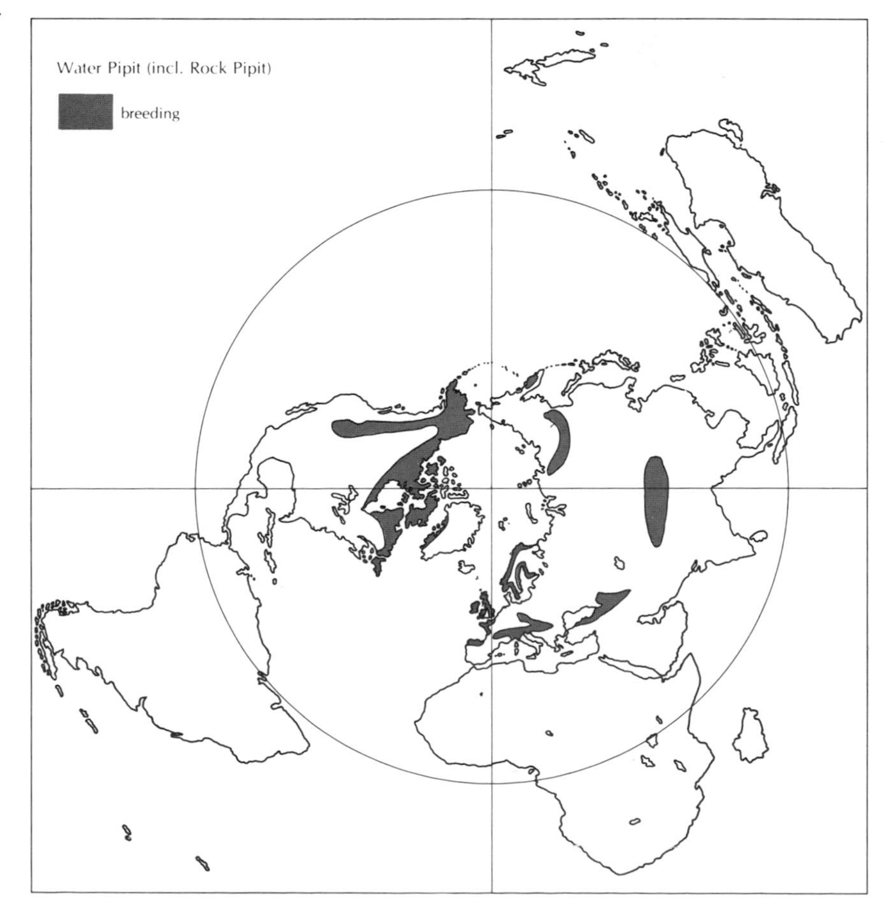

The breeding range of the Water Pipit (including the Rock Pipit).

Opposite: The Red-throated Pipit places its nest on the ground concealed under a tussock or low bush. It breeds right across northern Eurasia but there is only one recorded breeding in North America.

Opposite: The Tree Sparrow of North America has a superficial resemblance to the European Tree Sparrow but it is not closely related.

numbers of young in the nest. The timing of the spell of bad conditions was critical; a little earlier or later and it might not have had nearly the same effect. In the area under observation fourteen nests were located in which 69 eggs were laid. Of these, 57 hatched, but the number that fledged was only 15. The minimum air temperature during the bad weather dropped to as low as 2°C (35°F), instead of the more usual 5°C to 10°C (40°F to 50°F), and in addition there were strong winds, gusting to gale force, which would have had an added cooling effect. Contrary to expectation, the birds that survived were the youngest, the reason being that older birds require more food. This example relates just to the Water Pipits. None of the other passerine species in the area suffered such heavy losses, presumably because the stage they had reached in their breeding cycles was slightly different and therefore not so critical.

Yellow Wagtail
Motacilla flava
No less than eighteen subspecies of Yellow Wagtail have been described and named, and between them they breed throughout Europe, much of North Africa, and in most of Asia except for the extreme east and central parts. Only three of these many races are relevant here. First *thunbergi* which breeds in northern Scandinavia and through north-western Siberia. It migrates through central Europe and the Mediterranean basin to winter in tropical Africa and perhaps also India. Secondly *plexa* which breeds in Siberia to the east of *thunbergi* eastwards to the River Kolyma. It probably winters in India. Thirdly *tschutschensis* which breeds in northeastern Siberia and the western parts of Alaska. Its wintering quarters are in the East Indies. All three of these arctic races breed in grass and bog areas, placing their well concealed nests in tussocks. The clutch of five

There are many different races of the Yellow Wagtail, between them breeding throughout most of Europe and Asia including the low arctic.

or six eggs is incubated mainly by the female. Both sexes rear the young, which leave the nest at about thirteen days but do not fly for a further four or five. In their low arctic and subarctic habitat, the birds are rarely unable to obtain their almost exclusively insect diet.

Yellow-headed Wagtail [Citrine Wagtail]
Motacilla citreola
The Yellow-headed Wagtail is a close relative of the preceding species. It has a breeding range extending from the arctic coast of central Siberia and around the Kanin Peninsula south through central Asia to the mountainous regions of the Altai and Tibet, where it breeds at heights up to 13,000 feet above sea level. In the low arctic the bird breeds in an area where the Yellow Wagtail does not occur, and it appears to replace it almost exactly, with the same habitat and food requirements. The breeding habits are almost unknown: the average clutch is said to be four or five eggs, but neither the incubation nor fledging periods have been recorded, though they are unlikely to be very different from the Yellow Wagtail's. Its wintering areas are in India, Indochina, and China.

White Wagtail including **Pied Wagtail**
Motacilla alba
Like other wagtails, this one has evolved into a number of well-marked subspecies inhabiting separate parts of its vast range, which extends through Europe and Asia, with the exception of India and most of southeast Asia. It breeds right on the arctic coast of Eurasia, including some of the islands such as southern Novaya Zemlya, is also common in Iceland, and a small population breeds in southeast Greenland. The birds breeding in the British Isles form one of the subspecies, *yarrellii*, and are called Pied Wagtails. The habitats used by the White Wagtail are very varied over the whole of its range but in the arctic it is exclusively a waterside bird, seeking its insect food close to small pools and tundra marshes. Its breeding habits are little different from the Yellow Wagtail except for a liking of holes for its nest. This is most often placed in a natural cavity among rocks or a stream bank, but the species is often associated with man and will use holes in buildings instead. In winter the birds migrate south to the temperate zone, where they form communal roosts in reed beds and thickets. However the Pied Wagtail in Britain has taken to more artificial sites, including trees in city centres, warm pipes in power stations, and the insides of heated glass houses. The benefits from this way of defeating the winter weather are obvious.

Pechora Pipit
Anthus gustavi
Meadow Pipit
Anthus pratensis
The first of these two essentially boreal pipits breeds from central Siberia east to the Pacific coast, in scrub or wooded areas bordering the tundra, and it winters in southeast Asia. The

Pied Wagtails nest in rock crevices, always close to water, where they can find their predominantly insect food.

The Meadow Pipit is essentially a scrub and heath bird, but has spread into the low arctic in much of Europe including Iceland.

Meadow Pipit is its replacement to the west, breeding all over northern Europe, south into the temperate zone, and also north into the low arctic in Iceland and along the coast of southeast Greenland. In winter it migrates to North Africa. Its breeding biology is well known, but the Pechora Pipit has been little studied, though it is unlikely to differ much. The nest is always well concealed in a clump of vegetation. The usual clutch is three to five eggs which are incubated by the female for about fourteen days, and about twelve to fourteen days later, the young fledge.

Wheatears and Thrushes

There is one arctic wheatear and two low arctic or subarctic thrushes in this group. Where the preceding birds fed mainly on small insects, many of them aerial, these species eat ground-living insects, searching for them among vegetation or among ground litter. In addition the Fieldfare and Redwing eat considerable quantities of berries and other fruit in the autumn and winter. The Wheatear nests mainly in crevices in rocks and cavities in the ground, which are plentiful in the arctic, but the two thrushes prefer nesting in scrub or trees, though both occasionally nest on the ground. This preference, among others, may be the limiting factor in their spread into the arctic.

Wheatear
Oenanthe oenanthe

There are a number of races of Wheatear, the two principal ones being *leucorrhoa* and *oenanthe*. The former breeds in northeastern Canada, Greenland, Iceland, the Faroes, and winters in tropical West Africa. The fact that the Canadian birds cross the North Atlantic to winter in Africa demonstrates that they have colonised the area from the east. The second race breeds almost throughout Europe and much of Asia, except for the south and

Wheatears feed mainly on insects but can take buds and seeds when insects are in short supply.

southeast, right across to the Pacific Ocean, and into Alaska. It reaches the arctic along the north coast of western Siberia, but for the most part breeds to the south. Nearly all the birds winter in tropical Africa, even those breeding in Alaska, showing how the colonisation of the North American continent has taken place from the west.

Wheatears make their nest in a concealed rock crevice, in rocky holes in the ground made by animals, and in a variety of artificial holes and cavities. In the more fertile parts of the arctic they are quite common; for example in West Greenland as many as a dozen pairs were located in about 400 acres of tundra, with the birds breeding in scattered outcrops of rock. The clutch is between five and seven eggs, which the female incubates for about two weeks, and the young are fed by both parents and fledge in fifteen days.

An interesting observation was made in Baffin Island, Canada, concerning Wheatears' nests, but which may well be true of other passerine species too. It was found that the birds were reusing previous years' nests, a habit commonplace among geese, for example, but relatively little recorded among other arctic breeding species. One which was examined was composed of no less than eight distinct layers, each representing a breeding season, when a thin layer of new material was added to the old. This reuse has a two-fold

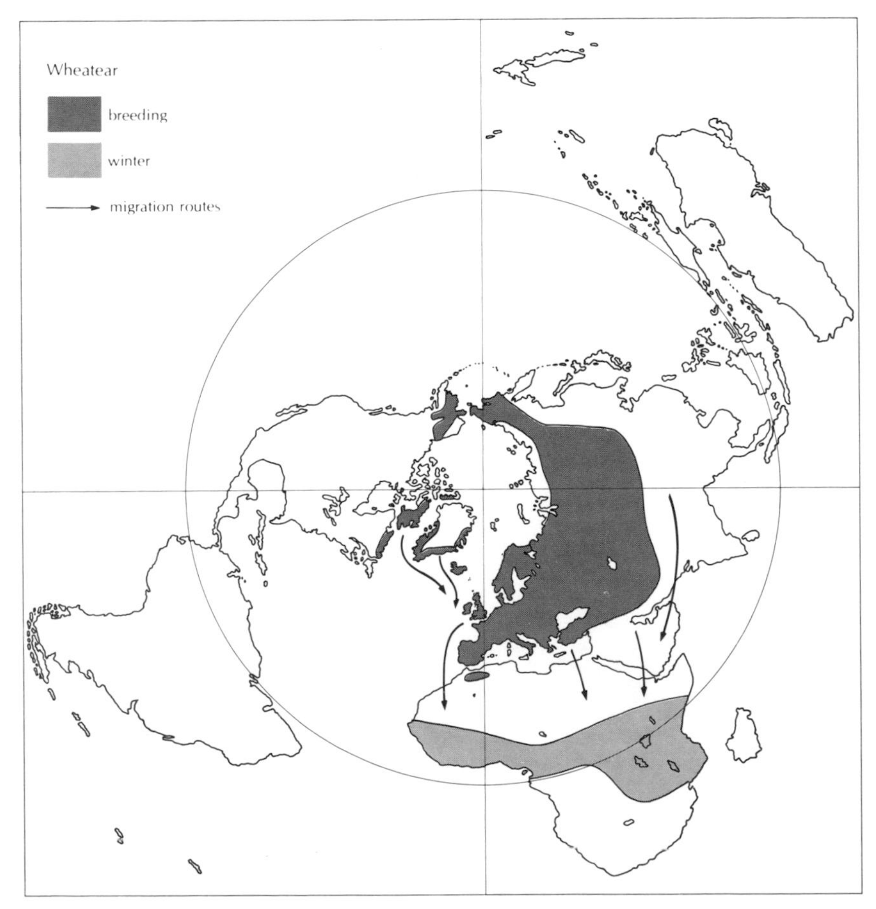

The breeding and winter ranges of the Wheater, together with its main migration routes.

function. First if the parents have successfully reared a brood in one year there is a strong probability that they will be able to do so again, as the site has demonstrated its ability to remain undetected by predators. Secondly, by knowing where they are going to breed, and merely having to refurbish an existing old nest rather than build a new one from scratch, the pair can save valuable days and thus lay their eggs that much earlier, for it is important that the chicks fledge before the end of the insect season.

Insects form their principal food in temperate regions but in the arctic they have a more mixed diet. However they cannot eat an entirely vegetable diet, clearly demonstrated by the way their numbers decline from the subarctic north through the low arctic to the high arctic. If the relative abundance of Wheatears in southwest Greenland, a fertile low arctic or even subarctic area, is taken as the standard, there are only one-third as many in a truly low arctic region further north, only one-twentieth as many in the southern high arctic, and in the northern high arctic they are virtually absent. In contrast, a predominantly seed-eating and plant-eating small bird such as the Snow Bunting is almost equally abundant throughout all four regions.

A colony of breeding Fieldfares was founded in southwest Greenland after a gale had blown some birds there in January 1937.

Fieldfare

Turdus pilaris

The Fieldfare is a typically boreal species breeding in trees and bushes, though occasionally on the ground. It has spread into the arctic in the last fifty years—partly by migration, partly by a freak circumstance of the weather—and then almost died out again, and is an interesting example of how colonisation of new areas can take place. In Iceland these birds have gradually become more common since the 1920s. Small numbers began to appear in winter in two or three localities, and it was assumed that these were migrants from northern Scandinavia, the nearest breeding locality. In the last twenty years a few pairs have even stayed behind and bred in the one area where there are some scrubby trees. This gradual colonisation has been attributed to the steady warming of the climate that has taken place there since the turn of the century.

By contrast the colonisation of Greenland happened virtually overnight and owes little directly to climatic amelioration. It was a scarce vagrant in the country until January 1937 when there was a sudden influx into the southwest corner. At the same time there were reports of birds from Jan Mayen Island and from northeast Greenland, and it seems from this that a flock or flocks were crossing the North Sea, probably on their way from Norway to Britain, when they got caught in a gale and were swept hundreds of miles north and west. The flock that reached south-west Greenland was fortunate as the birds found relatively sheltered valleys with thick scrub, berried plants, and quite passable Fieldfare habitat. Other flocks must have perished on reaching less favourable conditions.

It was not until 1944 that breeding in this area was actually first proved, but it seems certain that they bred right from 1937. The colony flourished and spread so that by the 1950s it was well established. However in the winter of 1966/1967 it was apparently completely, or nearly completely, wiped out when there were abnormally heavy snowfalls which covered all the bushes and other food plants. Since that time no Fieldfares have been seen in the original area though a few pairs may be hanging on further north, to where they had earlier spread. A remarkable and inexplicable feature of these colonists is that they lost their migratory instincts and became sedentary.

Redwing
Turdus iliacus
The Redwing is virtually confined to the boreal zone for breeding but in a few areas penetrates both the temperate and arctic. The latter is reached in Iceland where a separate

subspecies breeds, and in a few places in northwest Russia. The Icelandic Redwing winters in the British Isles and adjacent parts of the continent, while those from Scandinavia and Russia winter mainly in central and southern Europe to the Mediterranean, though some also come to Britain and northwest Europe.

The nest is a fairly bulky structure of twigs and grasses, though without any of the binding mud that is a feature of other thrush nests, and is normally placed in the fork of a tree or the crown of a low shrub. Nests on the ground are not uncommon, and here they are usually in a stream bank, or sometimes in an old tree stump. The clutch of four or five eggs is incubated by the female for about a fortnight, and the young fledge in another two weeks. Redwing occasionally breed in small colonies, while in Iceland they have become garden birds, nesting in shrubs and trees of towns and villages in a country where natural vegetation of this kind is uncommon. Their weak fluting song can be heard from the rooftops, or other vantage points.

Redpolls

Redpolls are small finches whose main food consists of the seeds of various trees, including birch, alder, larch, and various conifers, as well as those of smaller plants. In spring and summer they eat plant buds and insects,

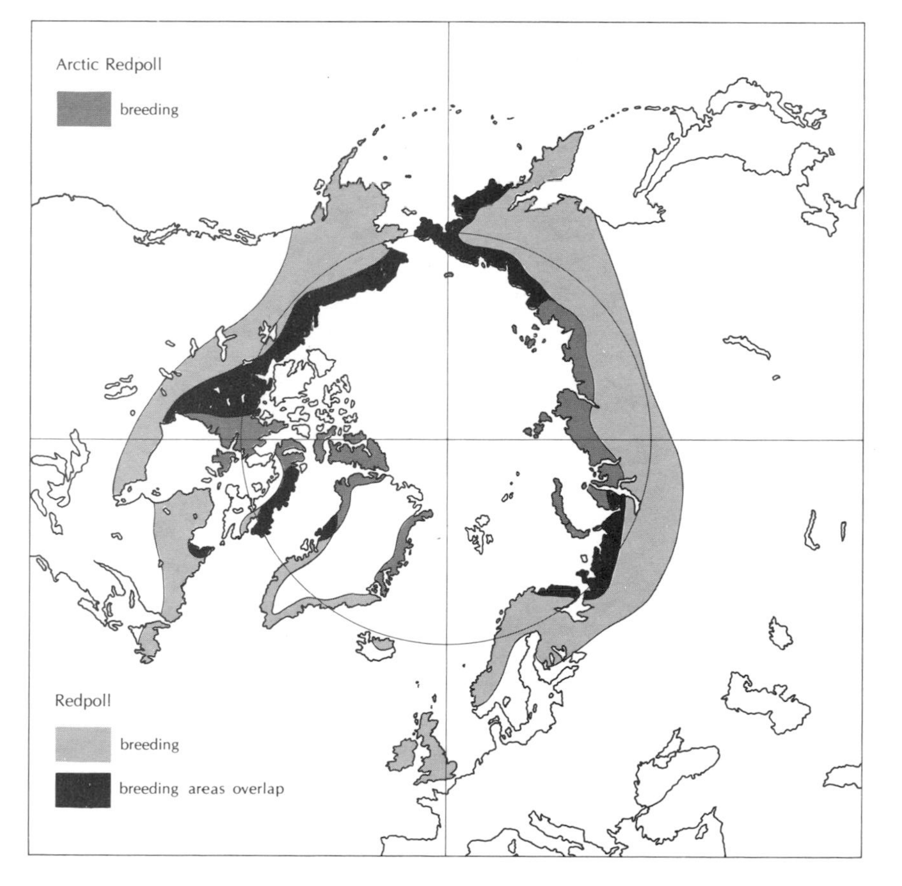

Opposite: The Redwing breeds in the arctic in Iceland and in parts of Russia. A preference for tree nests probably inhibits spread further northwards.

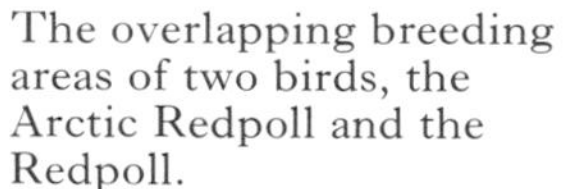
The overlapping breeding areas of two birds, the Arctic Redpoll and the Redpoll.

which they also feed to their young. However, reliance on insects is small, enabling them to live in the high arctic and even to winter in parts of it. Their taxonomic status is very confused. There are probably two species, one with four and one with two subspecies. One of them, the Redpoll, has a circumpolar distribution in the boreal zone, including the temperate zone in Europe, and also occurs in the low arctic zone in Iceland and southern Greenland, and on the mainland of arctic North America. The other, the Arctic Redpoll, is exclusively arctic, breeding in the high arctic in Greenland, Canada, and parts of Siberia, and also in the low arctic of Scandinavia and North America where it overlaps with the Redpoll. In Scandinavia the two species interbreed freely, suggesting that they are only subspecies, but in North America, although they live side by side over wide areas, they do not apparently ever hybridise and so qualify for treatment as separate species.

Redpoll
Acanthis flammea

There are four subspecies of Redpoll of which only one, *cabaret* or the Lesser Redpoll, does not reach the arctic, being confined to the British Isles and some areas of continental Europe. The other three are: *flammea*, the Mealy Redpoll, which breeds from northern Scandinavia across Russia to the Pacific coast and also through Alaska and northern mainland Canada, though not on any of the islands; *rostrata*, the Greater Redpoll, which breeds in the southern half of Greenland on both east and west coasts and also on Baffin Island; and *islandica*, which is confined to Iceland. This last race is completely sedentary but the other three all move south in winter into the boreal and temperate zones.

Redpolls breed in open forest, scrub, and marshy areas with low willow and alders. In the arctic they are more or less confined to regions where there is still low scrub, though small numbers are found in more typically tundra areas where the nest is actually on the ground in a tussock, or perhaps in a rock crevice. The clutch of four eggs, occasionally five, is incubated by the female for eleven to thirteen days, and the fledging period is about thirteen days. In winter they gather in flocks, often numbering several hundreds, and feed on seeds, particularly of alders and larches. Some localities have a tradition of being the winter haunt of these birds and ringing has shown that many return to the same site year after year.

Two species of Redpoll breed in the arctic but their relationship in areas of overlap is unclear.

Arctic Redpoll
Acanthis hornemanni
The Arctic Redpoll has two subspecies: *hornemanni*, sometimes called Hornemann's Redpoll, which breeds in the high arctic on the east and west coasts of Greenland and in northeastern Canada, particularly on Ellesmere, Bylot, and north Baffin Island; and *exilipes* or Coue's Redpoll, which breeds across the Eurasian arctic from Scandinavia to northeast Siberia, in north Alaska, and on the northern fringe of mainland Canada. Most of the latter subspecies move south short distances for the winter and adopt similar winter habits to the Redpoll. However the birds of Alaska and parts of mainland Canada are sedentary, as are virtually all the Hornemann's Redpolls, making them the most northerly wintering passerines. Only in the very severest conditions will they move.

These hardy small birds have adapted in a number of ways to a high arctic life, particularly necessary in winter. The most important adaptation is that their oesophagus, the rather narrow tube connecting their throat to their stomach, is enlarged to form what is in effect a crop, something which birds of this family do not normally have. It allows them to take in more food than they could otherwise manage during their very short feeding period in winter, and to digest its contents at leisure through the night. They further have the ability to increase at low temperatures their digesting efficiency when it would normally tend to fall off. Their main food in winter is seeds, especially birch, which have a relatively high calorific value, and observation has shown that they are capable of extending their limited feeding period each day by carrying on eating in conditions of very poor light. Finally the birds in the high arctic have a much paler plumage for maximum heat conservation than those to the south, even though the pattern is similar.

Their breeding habits are also adapted to a life in a colder climate. In the majority of birds, incubation of a clutch of eggs begins after the last has been laid so that they hatch more or less simultaneously. Exceptions include such birds as raptors and owls and the Arctic Redpoll joins these in starting incubation with the first egg, not because it will only rear the oldest youngster in a year of poor food like the owls, but to stop it from freezing. Hatching is thus asynchronous, but rearing success seems not to be affected. The incubation period for one egg and the fledging period of the young are both about eleven days and so a day or two shorter than the Redpoll's. The birds manage to winter in high arctic Greenland by retreating not to the coast, which further south might be the warmest part of the country, but by going inland into some of the valleys or even hill plateaus where the winds sweeping off the icecap prevent heavy snowfalls, thus always leaving seed-bearing scrub and other plants exposed for the birds to feed on.

Sparrows

These North American sparrows are more closely related to the buntings than to Eurasian sparrows. There are four species in this group. Two, the Savannah and the White-crowned, are widely found in the low arctic, while the other two, the Tree and Harris's, only just reach it. All four are primarily seed eaters, but take insects at the height of their abundance in summer and feed them to their young in the nest. None of them penetrate the high arctic.

Savannah Sparrow
Passerculus sandwichensis
The Savannah Sparrow is a very widespread North American species occurring almost throughout Canada, except the extreme north, and the United States, except the southernmost states. It has been divided into a large number of subspecies, based on small variations in size and colour. Two or three of them reach the Canadian low arctic and Alaska, breeding almost to the limits of the mainland coast but not on the arctic islands. They are

birds of open habitat, including grassland, marshes, and vegetated sand-dunes, and they live in scrubby areas too but not among trees. In the arctic this can be translated into scrub tundra and to a lesser extent open tundra with only low vegetation. Their nest is almost always well concealed on the ground, sometimes completely enclosed in the bent over grass stems remaining from the previous year. The clutch is usually five or six eggs which are incubated for twelve days by both parents in about equal spells, and the young fledge in about two weeks. If disturbed from the nest the adult will go into a distraction display, fluttering low over the ground pretending it cannot fly. Indeed their flight always appears rather weak and fluttering, that is if they can be persuaded to fly at all, for they are as likely to run quickly away, low and scurrying, a little like a mouse, as they are to fly. And when they do take off it is only to drop into cover again within a short distance. Despite this they are quite long distance migrants, wintering well to the south of their breeding grounds, around the Gulf coast of the United States and through Mexico and much of central America.

White-crowned Sparrow
Zonotrichia leucophrys
The present distribution of this species suggests that it was once confined to the

The White-crowned Sparrow's breeding and winter ranges.

The White-crowned Sparrow breeds widely in North America but was probably once confined to the low arctic, from which it has spread south.

subarctic and low arctic, from where it has spread out. It now breeds throughout the northern mainland of Canada and most of Alaska, as well as down the west side of the United States through California, Nevada, Arizona, and into New Mexico. In these more southerly areas it probably spread via the mountain ranges which would provide a somewhat similar habitat to the arctic, and it has only comparatively recently adapted to living in warmer climates and colonised these states more widely. It winters throughout much of southern and western North America.

The large range of the White-crowned Sparrow is occupied by a number of subspecies varying slightly in plumage characters, particularly in the completeness or otherwise of the white stripe above the eye. A further difference is that their song varies geographically. It has the same basic notes and shape throughout the range, but birds living in different localities have their own, often quite constant, variations on the theme. These are the equivalent of dialects and a really experienced person could identify where he was in northern Canada by carefully listening to the sparrows singing. Very occasionally a bird with the 'wrong' dialect appears in a locality and sings in competition with the other birds. The breeding habits of the White-crowned Sparrow differ only a little from the Savannah Sparrow. The nest is a rather more bulky affair, incorporating twigs and bark as well as grass, and is lined with finer material. The usual clutch is four or five eggs, incubated by the female only for between twelve and fourteen days, and the fledging period is about two weeks.

Tree Sparrow
Spizella arborea
Harris's Sparrow
Zonotrichia querula

The first of these two subarctic species must not be confused with the European Tree Sparrow of which it is no relation. This one may well have been named by early settlers who saw in it a superficial resemblance to the European bird but its true habitat is scrub and scattered small trees in the zone between the true forest and the tundra, often called the muskeg. Its breeding range extends from western Alaska across northern Canada to Labrador and Newfoundland, though only touching the arctic coast in northwest Canada, and it winters in a broad belt across the middle of the United States from California to North Carolina, favouring open fields and other weedy places.

Harris's Sparrow has a much more restricted breeding range, occurring only in a fairly narrow belt from the western coast of Hudson Bay north and west to the arctic coast in the Mackenzie District. It winters in west and south United States from Utah and California to Texas and Tennessee, seeking hedges and the borders of woods. Its breeding biology has been little studied; indeed a nest and eggs were not discovered until 1931, after a long and tedious search made much harder by the bird's secretive habits, being very difficult to follow back to its nest and equally hard to flush from it. The usual clutch is four eggs and the female incubates for thirteen or fourteen days. Both parents tend the young but the fledging period has not been recorded.

In contrast the Tree Sparrow was the subject of a very detailed breeding study in the 1930s. This revealed some interesting data, for the complete record of the time spent in the various stages of the breeding cycle was discovered, and it probably differs little for a great many similar sized species in the sub-arctic and low arctic. After the birds arrive in flocks they almost immediately break up and pair. Nest building is started at once and lasts for an average of seven days. There follows a rest period of two or three days before laying is begun, which takes as many days as there are eggs, usually between four and six. Incubation begins as soon as the last egg is laid and occupies the female for the next twelve or thirteen days. She sits for about two-thirds of the total time, spending the other third off the nest feeding, but the spells on and off alternate at quite short intervals so the eggs have no chance of cooling too much. Hatching takes place on the same day for all the eggs, often within a twelve-hour period. Both parents feed the chicks, though the female spends more of her time in the first few days brooding the young. They leave the nest at about nine days old, a few days before they can fly, and they may disperse up to 100 feet away but are still fed by their parents. Indeed this parental feeding continues for a further one or two weeks after they have fledged, completely inhibiting any possibility of a second clutch. Both adults and young now undergo a moult which last about three weeks. Finally, all the birds from a wide area gather and remain in flocks until they migrate after about a month. Species breeding further north will be able to shorten some aspects of this four-month period, for example the moult and the period spent feeding in flocks before migration, in order to fit their breeding cycle into the shorter summer. Another saving would be to reuse the previous year's nest, as does the Wheatear.

Buntings (Longspurs)

There are three species of buntings breeding in the arctic. Two, the Snow Bunting and the Lapland Bunting, have complete circumpolar distributions, while Smith's Longspur is restricted to a comparatively small area of Alaska and Canada. The buntings, like the sparrows of the previous section, are predominantly seed eaters, but take insects in times of plenty and feed them to their young. The Snow and Lapland Buntings, particularly the latter, are among the commonest of the small birds of the arctic.

Lapland Bunting [Lapland Longspur]
Calcarius lapponicus
There is considerable overlap in range between this species and the Snow Bunting. Both are completely circumpolar, the Lapland Bunting having a small gap in Iceland. The Snow Bunting breeds very much further north but their southern boundaries are very close together. Neither penetrates the boreal zone but both breed on mountain ranges, such as those forming the spine of Scandinavia or the Kamchatka Peninsula, which bring them well to the south of the arctic. The Snow Bunting also breeds in very small numbers in northern Scotland. With two closely related species having such great overlap it must clearly be a habitat difference that prevents them competing—whereas the Lapland Bunting prefers well vegetated tundra, including marshes, heaths, and even low scrub, the Snow Bunting is much more at home in the barren areas. In winter Lapland Buntings migrate to temperate latitudes, being particularly common on coasts and wide open areas such as large fields, airports, and estuaries.

They are not colonial birds in their breeding but quite high densities are reached in some areas. On Bylot Island in Canada, for example, several times nests were found only 100 yards apart, and there were up to ten nests in one four acre plot. In another part of the Canadian arctic, on Baffin Island, densities were not so great but numbers were nevertheless high, taking into account the very large areas of relatively uniform habitat. One area of marshland of about 400 acres had as many as 170 breeding pairs. This compared with 20 pairs of Snow Bunting, the next commonest species in that area, although it too reached comparable densities in more typical, drier habitat.

The female builds the nest, and after laying the clutch of five or six eggs, she incubates them for about twelve to fourteen days. The male then helps her feed the brood, which leave the nest at about eight or nine days old, some three or four days before they can actually fly. In the more northerly areas the female may start to incubate with the first egg to protect them from the cold. The amount of time the adults spend feeding the young also differs with latitude. In the most southerly areas, at about 55°N, the parents have a night time rest period lasting about seven hours. In the far north, at about 76°N, they only stop for between three and five hours, thus making maximum use of the continuous daylight.

Male Lapland Buntings help their mates to feed the breed but the females have to do all the incubation themselves.

Smith's Longspur
Calcarius pictus
A relatively little known species, Smith's Longspur occurs in a restricted belt of low arctic from the southern shore of Hudson Bay northwest to the Mackenzie District and northern Alaska, and it winters in open grass areas in the southern United States. It breeds in well-vegetated tundra, including some scrub areas. The nest itself is usually on, or rather in, a tussock, and is built of grasses lined with finer material, including feathers. Usually four to six eggs are laid, which are incubated probably by the female alone for eleven or twelve days. The fledging period has not been determined but is unlikely to be much different from the twelve days of the Lapland Bunting.

Snow Bunting
Plectrophenax nivalis
Without doubt this is the most northerly breeding passerine bird, and not just sporadically but regularly in all the highest latitude lands round the top of the globe. Its complete circumpolar range includes all the island groups to the north of Eurasia and North America, and in addition it extends south down mountain ranges. Both on the mountains and the tundra its habitat is barren rocky areas with little vegetation, and it also breeds on rocky shores, sea cliffs, and even nunataks sticking out of inland icefields. Migration does not take the Snow Bunting very far south, and certainly not far from ice and snow, though the majority do leave the

The Snow Bunting's summer and winter ranges.

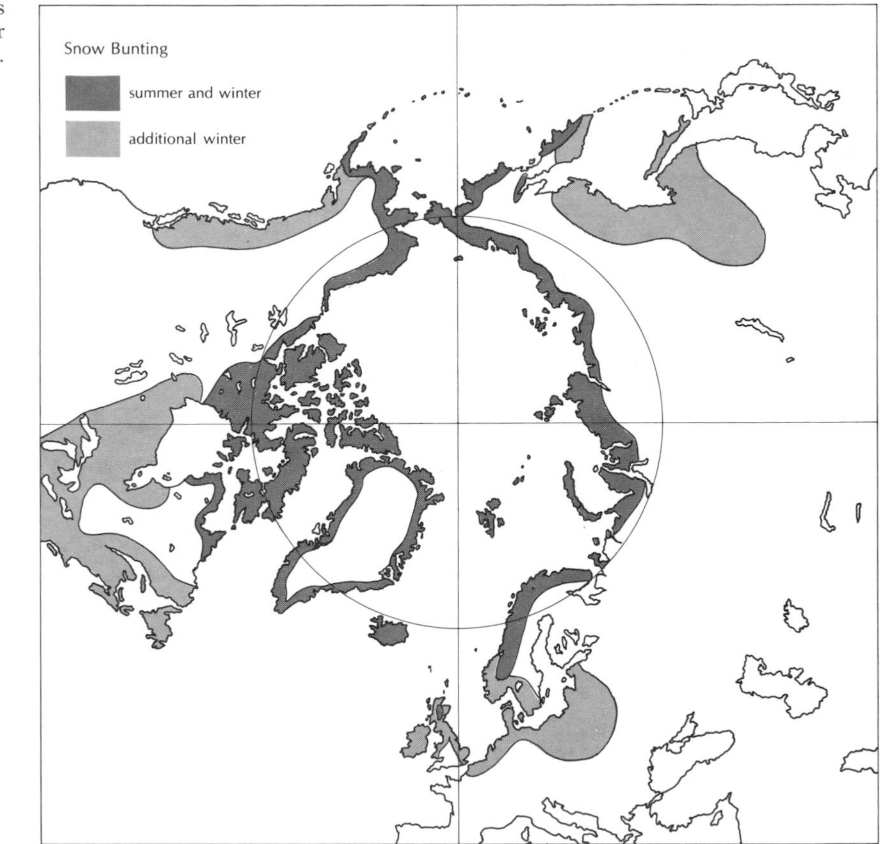

high arctic. The bird winters in cold areas and is nearly as hardy in this respect as the Arctic Redpoll. Bleak shores and open fields are its favoured terrain. However in some areas it chooses to live in the centres of villages and even towns, becoming like the House Sparrow in its tameness and dependence on man for food scraps. Even away from areas such as these it is rarely concerned about man's presence, whether on the breeding grounds or in a flock encountered on some windswept and wintry shore.

The birds breed in rock crevices, often deep inside, and nearly always out of sight. Here the female builds the nest of dead grass stems, leaves, and moss, lining it with finer material including hair and feathers. It takes her about four days to build, rather quicker than the low arctic Tree Sparrow. She incubates the clutch of four to seven eggs alone, but is often fed on the nest by the male. Both incubation and fledging periods last about twelve to fourteen days. The young sometimes leave the nest a few days before they can fly, but being well concealed in their rock crevice site there is less to be gained by early departure than there is for a ground nesting species. The nests of the very few pairs of Snow Buntings breeding on mountains in northern Scotland are notoriously difficult to find. Nothing, though, could be easier than finding them in the arctic. If the bird does not actually lead you straight there, a very few minutes watching will suffice. Alternatively in some areas it is possible to spot where the nest is most likely to be, because there are many areas of tundra where the only features are occasional rock outcrops, and they invariably choose these to nest in.

The male has a strikingly contrasting black and white breeding plumage that he shows off to best advantage in a slow, half gliding display which is reminiscent of a large black and white butterfly flitting over the tundra. His winter plumage is more subdued but is still essentially pale, like the female's, as a protection against the cold. There are a number of races of Snow Bunting, the palest being found in northeastern Siberia, not because it is the snowiest part of the arctic but because it is the coldest. The moult of the Snow Bunting takes place immediately after the end of the breeding season, and in order to speed up the process the adults will sometimes shed as many as four or five main wing feathers at a time instead of the more usual one or two. Their powers of flight can become considerably impaired for a week or so before their new wing feathers have grown, but for a ground-living species not requiring skill at flying to catch its food, this is a small price to pay. The change back from the dull winter plumage to the breeding dress is accomplished without a moult; the duller edges of the feathers simply abrade away. Another area of abrasion, however, necessitates an extra if limited moult, which may be unique to Snow Buntings. During the early spring they are often feeding on snow, picking up seeds that have blown onto it. The hard frozen surface gradually wears away the feathers from around the bill and down their chin, but these are replaced with new feathers in spring, just before the start of breeding.

Crows

There is just one member of this family breeding in arctic. Most of those familiar in the boreal and temperate regions breed in trees, but the sole arctic representative, the Raven, prefers cliff ledges.

Raven

Corvus corax

The largest of the passerines, the Raven occurs almost throughout the whole of the northern hemisphere. It has a wide range of habitats including mountains, deserts, sea cliffs, forests, and also the arctic tundra. It breeds in the high arctic in Greenland, several of the Canadian arctic islands, north Alaska, and northeast Siberia. The bird is sedentary throughout most of its range, the arctic being no exception, and it shares with the Arctic Redpoll the distinction of wintering the furthest north.

Ravens build their bulky nests on cliff ledges, using sticks, if they can find them, seaweed, moss, and grass. The usual clutch is four to six eggs which the female incubates for three weeks. She does not begin incubation until the clutch is complete but in the far north half buries the earlier eggs in the lining of the nest to help protect them from the cold. Similarly the newly hatched chicks are covered by the nest lining to help insulate them. While the female is sitting she is regularly fed by the male, and this saves her getting off and thus allowing the eggs to cool. When the young are small they are continuously brooded by the female while the male collects food, but occasionally these roles are reversed. The whole breeding cycle is quite long and may restrict how far north the species can breed. On the other hand their cliff nesting site is always available early in the season, and they have a very catholic taste for food which also helps. Their diet is based partly on predation, partly on scavenging. The former includes eggs and young birds in summer, and rodents in winter. Scavenging takes place round man's settlements, but also includes eating the droppings, or at least parts of them, from caribou and seals, and anything edible they can find on the seashore.

Completely black plumage might not seem the most suitable for an arctic existence, but the Raven is quite a large bird and so can probably counteract this disadvantage with

The summer and winter ranges of the Raven.

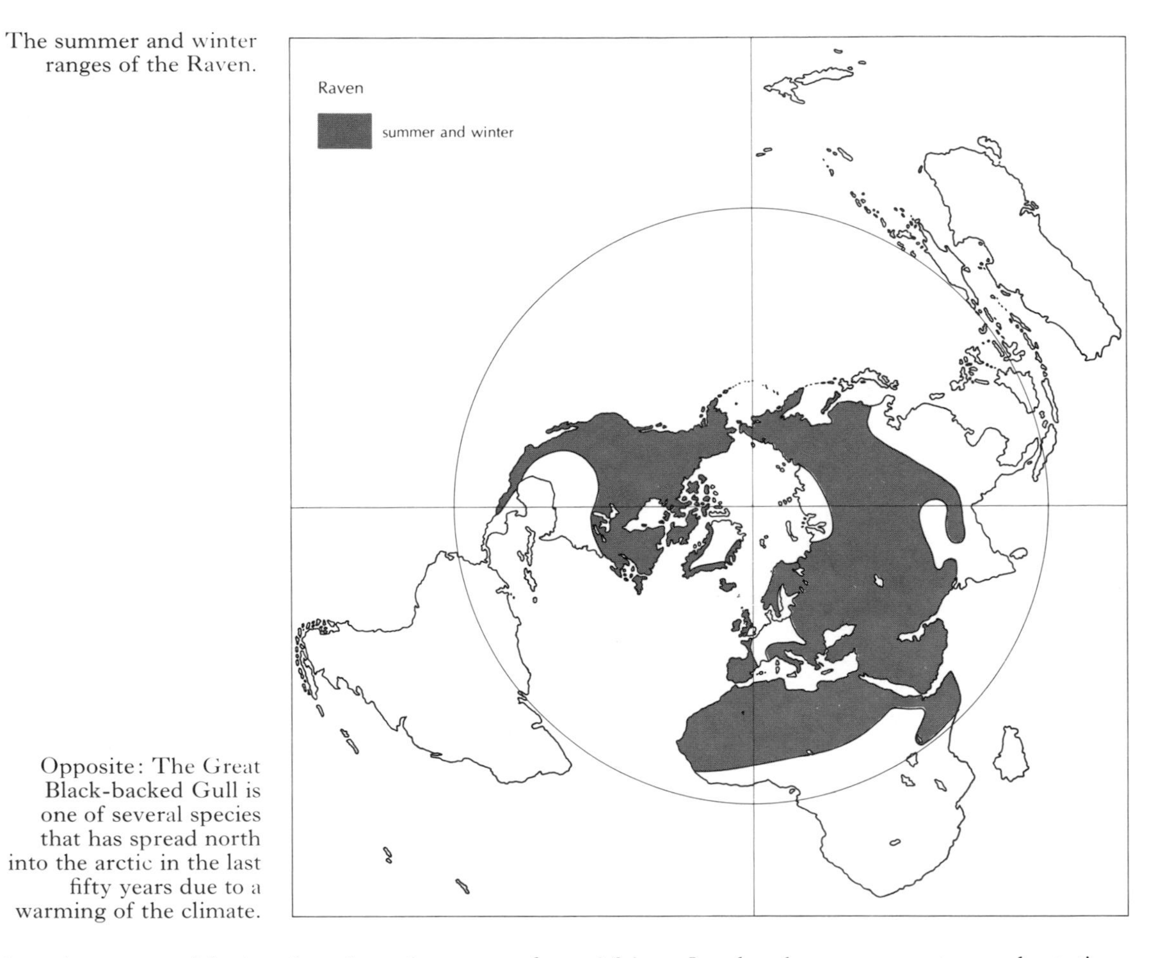

Opposite: The Great Black-backed Gull is one of several species that has spread north into the arctic in the last fifty years due to a warming of the climate.

the advantage of losing less heat because of its size—the average size of the arctic birds is larger than those further south, for exanple by about 15% compared with those in North Africa. It also has one or two adaptations including thick pads on the undersides of its feet, which do not occur on birds living further south.

7 The conservation of arctic birds

'The arctic circumpolar zone represents the largest continuous tract of wilderness and wildlife habitat now remaining in the northern hemisphere'

Sir Frank Fraser Darling 1969

The majority of birds that live in the arctic, and certainly all those in the high arctic, are at the limit of the conditions that will permit them to breed successfully. Indeed, as already mentioned, many species are unable to nest every year but suffer periodic complete or partial breeding failures as a result of adverse weather or a lack of sufficient food, or sometimes both. All the different birds have adapted in their various ways to this rather precarious existence but their continued survival depends on the maintenance of the status quo.

The fine balance of controlling factors can be upset in two ways: first by some alteration in natural conditions, and second by interference from outside. Natural changes are most likely to be brought about by variations in climate. Our detailed knowledge of the arctic and its bird life is comparatively recent and such climatic change as has occurred in the last hundred years or so has been towards amelioration, with significant rises in mean temperatures in many areas. However this warming has been more marked in winter than in summer, and more noticeable in its effect in the sea than on land. Consequently although there have been some dramatic northward shifts in fish populations, changes in the numbers and distribution of birds from this cause have been much less important. Certainly some species have spread north, for example the Wheatear, Pintail, and Great Black-backed Gull, while the most southerly breeding outposts of the Snow Bunting and Turnstone are in decline, but these are negligible compared with their total distribution. Other effects of this change on bird life have included the more frequent appearance of southerly species as stragglers, perhaps the prelude to colonisation if the warming continues, and the slight, but perhaps important, lengthening of the possible breeding season in some places.

In the last few years there has been much talk of a reversal of this arctic warming but the trend, if it is one, has not been established for very long, and so far there have been no corresponding changes in bird numbers or distribution that can be attributed to it. However, there has been an instance of a delayed migration northwards: Snow Geese passing through Manitoba are now averaging four to six days later than they did in a fifteen year period ending in 1937. It is suggested that the geese are delaying their arrival on their breeding grounds on Baffin and Southampton Islands because the lower temperatures recorded there have shortened the breeding season, but there may be other factors at work and further evidence is needed. In any case alterations in climate are essentially long term and although they can

Eskimos kill hundreds of thousands of Brunnich's Guillemots in West Greenland each year. The present population can probably withstand this harvest provided disturbance at colonies is avoided.

be detected over periods of only a few years, a much greater length of time is usually needed before the effects can be measured.

Outside interference simply means man and his activities. The Eskimo has been living in some parts of the arctic for hundreds of years. The relationship between him and his environment has now changed out of all recognition, but for centuries there was complete dependence on natural resources and no succour available from outside if those resources failed. There is no record that the Eskimos or the northern Siberian tribes ever practised conservation. When animals or birds were abundant they killed and ate as many as they wanted, both for immediate needs and to store for the winter. If hunting was bad they either moved to better areas or starved to death. No one taught them that over-harvesting could contribute to the disappearance of their quarry. Precise proof is lacking but the most probable cause of some groups of Eskimos dying out is starvation following poor hunting seasons.

Eskimos once lived all along the northeast coast of Greenland from Scoresby Sound northwards to at least 80°N. The only ones actually to be seen there by a white man were a small party in 1823. Present day Eskimos at Scoresby Sound, brought there in 1924 from Angmagssalik 400 miles south down the coast, live in wooden buildings and have many of the trappings of modern civilisation, but along the shores of the great fjords of northeast Greenland one frequently comes across the circles of stones marking their summer camps of a hundred or more years ago. Excavations have shown that these people had dogs and hunted the caribou or reindeer, among other animals. It is certainly more than just coincidence that the caribou, too, is now extinct in this region. Formerly it was both numerous and widespread and to this day the cast antlers can be found lying on the tundra, many looking as if they are only a year or two old. Yet the last sighting of a caribou there was in 1899. Whether hunting alone or a combination of that and adverse weather conditions caused them to die out is not known, but the evidence points to a near simultaneous extinction of the Eskimos too.

In many areas the wildlife has flourished despite regular depredations by the Eskimos. The latter's numbers were never great and not unnaturally they concentrated on those species which were the most numerous and so best able to stand repeated culling. The great seabird colonies would barely notice the taking of even many tens of thousands of their teeming inhabitants, while some populations of wildfowl were, and indeed still are, able to tolerate the regular killing of comparatively large numbers. It was the introduction of the rifle and shotgun that brought about a change for the worse, that and an increasing population. Modern principles of conservation are grasped only slowly by an Eskimo and enforcing controls on hunting is well-nigh impossible. The last few decades have seen a steady improvement in the situation though some of the progress has been offset by the great increase in hunting, often indiscriminate, by men working at mineral exploration camps and other installations. However there are now few species that are directly threatened, at any rate by the indigenous inhabitants of the arctic, though there remains a legacy in some of the more populated areas where bird and animal life has been sadly depleted and has not yet recovered.

Although all Eskimo settlements now have the benefit of a store where they may buy food in cans, and there are many families who have swapped the snowdrift outside the door for the modern freezer, killing of wildlife for food still takes place on a large scale. There are few figures available from the Soviet arctic, but the Danes in Greenland and the Americans and Canadians have carefully investigated the annual kill of many bird species so that they can if necessary give them some protection, as well as allowing for this kill when formulating hunting regulations for the migration and wintering areas.

The population of just under 10,000 Eskimos living in the Yukon–Kuskokwim

Delta area of western Alaska takes in a year about 5,500 Whistling Swans, just over 1,000 Sandhill Cranes, and nearly 40,000 eggs, mostly of ducks and geese. It is not thought that these figures are in any way excessive. The population of Whistling Swans is put at between 70,000 and 90,000, so the annual kill amounts to between 6% and 8%, well within the reproductive capacity of the species even allowing for other mortality. The Eskimos also kill large numbers of geese, including about 8,000 each of Emperor and Brant, 6,000 Snow Geese, and no less than 22,500 Whitefronts and 38,000 Canadas. Of these five species, the first four have populations numbering hundreds of thousands in that part of Alaska alone, and the Eskimo's kill amounts to between 2% and 11%, but the Canada Geese are virtually all of the same race called the Cackling Goose, of which there are only some 80,000 breeding there, so the kill is relatively a very high one.

Figures for the annual take of several bird species in western Greenland have come from a combination of kill statistics plus results from the national ringing or banding scheme. The most important species for the Greenlander in economic terms is the Brunnich's Guillemot. It is estimated that about 750,000 are shot each year. It forms a principal summer food for some communities, as well as being salted and dried for the winter. There is also a more limited amount of egg collection, amounting to several thousands a year, but the sheer inaccessibility of the ledges on the cliffs probably restricts this activity. The birds are shot either at the colonies or during the winter when large numbers are concentrated off the southwest coast. Shooting at the colonies is the most damaging as the general disturbance causes great loss of eggs and chicks, in addition to the birds actually killed. A number of colonies have now been placed 'out of bounds' for summer shooting, but such regulations are extremely difficult to enforce, and the Eskimo has little respect for them. The population of Brunnich's Guillemots in west Greenland is certainly numbered in millions, and it is probable that the annual kill at its present level is not too serious. However at least some of the birds taken in winter belong to other populations breeding in Spitsbergen and northwest Russia, and the effect on these must be considered. A further mortality of these birds takes place in fishing-nets and as many as 15,000 have been drowned in them in a year.

The other major food species in west Greenland is the Eider Duck. Large numbers are shot annually, thought to be around 145,000 in the early 1950s, while egg collecting is also carried out at several colonies. The latter is now severely restricted but the Eider has decreased markedly in west Greenland in the last fifty years, and though climatic factors may be partly responsible, over-hunting is the more obvious cause. There have been several recent regulations to try to protect it, in particular its breeding colonies, but it is too early to measure their effect.

Just as the indigenous tribes of the arctic killed the birds that lived there for food, so the people living further south have long been hunters too. They took adults, eggs, and young from seabird colonies, and eggs of all kinds of birds. However the greatest attraction was always the flocks of geese, ducks, and waders, that appeared in the autumn and departed the following spring. Only relatively small numbers of these kinds of birds breed in temperate lands; it was the wintering numbers that were so impressive, and so attractive to a hunter at a time of year when other meat was scarce. He devised many ingenious ways of catching and killing them, including flight and clap nets, and decoy ponds. The invention of a reasonably safe and portable gun brought many more kinds of birds within reach, and probably for the first time seriously threatened the existence of some arctic species. The Eskimo Curlew was slaughtered almost out of existence, while other birds, including the Hudsonian Godwit and several species of geese, have also suffered from gross over-shooting. And what makes it worse is that most of this was no longer a necessity to provide food for the winter—the idea of shooting for sport had taken hold.

Sport is now seen as providing much needed relaxation and it also brings considerable enjoyment to the participator, and in many cases to the spectator too. Killing wildlife for sport was coupled by our fathers and grandfathers with a degree of competitiveness which now is thankfully to be seen only in sporting games as opposed to sporting pursuits. Today the idea of creating new 'bag' records is nearly as extinct as the Eskimo Curlew whose downfall it was. However the decline of mass slaughter by the few has been replaced by an enormous growth in the numbers of hunters. For the sake of their quarry it is absolutely necessary to limit what they can shoot.

The belief in the conservation of wildlife as a resource is a comparatively recent phenomenon, but there is already a generation grown up who have known no other philosophy. There are still great areas of the world where the message has not yet penetrated, and may not for many years to come, which will be to the detriment of several kinds of wild animal and bird. But in Europe and North America at least, modern hunters are used to regulations and restrictions of their sport which they recognise as in the best interests of everyone. North America is far ahead of Europe in such matters. Their network of reserves, their management techniques, and the control of shooting through bag and cartridge limits effectively helps both hunter and hunted. International cooperation between Canada, the United States, and Mexico has, despite occasional disagreements, done nothing but good. International cooperation in Europe is more recent, but is now working well. Reserves are fewer in number and management less refined. Bag limits are rare but education of hunters has been excellent so that few shooters today are unaware of the need for self control. While no one tells them how many geese or duck they

A duck decoy showing two of the netting-covered 'pipes' in which the ducks are caught, formerly for eating, nowadays for ringing and release in conservation studies.

may shoot, cases of deliberate slaughter just because the chance was there are extremely uncommon today.

On both continents a majority of the quarry species of geese occur in the arctic. Here they are subject to wide fluctuations in breeding success which produce marked changes in their numbers, even in the short term. In North America a sudden decline is immediately reflected by reduced bag limits or even a total ban on shooting in their wintering areas. In Europe it so happens that the majority of goose populations have enjoyed a long period of prosperity that has brought about a steady increase in numbers over the last two decades. Statutory protection has been given to the rarer species, and in most cases these have responded with welcome increases. Hunters too have instituted local voluntary bans on shooting declining stocks, and the number of refuges has been growing in all countries.

There are, then, few arctic breeding quarry species that are threatened in their migration and wintering habitats, provided these lie in the temperate zone, the home of 'civilised' man. Where the birds occur in less conservation-conscious countries they are still at risk but even here the direct threats are not particularly serious. Of more concern is the continuing loss of winter habitat to building, drainage, reclamation, and other changes of use. Most geese have proved adaptable when faced with the disappearance of some major haunt, finding alternative sites and foods very quickly. Indeed the ability of many goose species to feed on farmland instead of natural vegetation has brought them into direct conflict with agricultural interests, a problem that has yet to be resolved to the satisfaction of all. Waders or shorebirds, on the other hand, are far more tied to their use of muddy estuaries and shores for winter living, and it is highly doubtful whether even a few of their number could survive on other habitats. Proposals to reclaim estuarine mudflats are constantly being made, and some have already been carried out. The displaced birds have no alternative but to concentrate on the remaining areas, where food resources may be being utilised to the full by other species. Starvation and death for at least some must be the only result.

Pollution in all its forms is now recognised as one of the biggest threats to all forms of life on this planet, including man. The persistent organochlorine pesticides are now gradually being withdrawn though not before their residues have been found in organs of birds from the arctic to the antarctic. Oil is perhaps the biggest identifiable menace at present. It affects more birds more widely than any other existing pollutant. In Europe and North America we have got used to recurring oil spills, mostly small, but occasionally very large, but bird life suffers immediately and very obviously. It is the biggest danger to seabirds, and can affect them both in summer when they are concentrated near their colonies, and in winter when they may be equally concentrated at some favoured feeding area. Oil cannot be banned like DDT. Oil cannot be kept away from seabird haunts because ocean currents and tides can drift it hundreds of miles. Oil can rarely be cleaned up again once it has been spilt. Oil is transported around the world in ships that are constantly exposed to the hazards of marine navigation. Man is seeking oil in the arctic, and has indeed already found it in large quantities.

The impact of man on the arctic, excluding the Eskimos and others already there, began in the sixteenth and seventeenth centuries when explorers started their searches northward for the hoped-for passages east or west to the Indies. It received an even greater boost from the desire for the products of the seal and the whale. Literally hundreds of ships and thousands of men went whaling in the seas around Greenland and Spitsbergen, and several shore stations were built where the animals could be processed. It is perhaps fortunate that these men were mostly too busy to bother much about the birdlife, although their desire for fresh eggs and meat often led to devastating raids on Eider and goose colonies.

Once over-greedy man had successfully exterminated most of the whales, and thus brought to an end commercial whaling in these waters, the wild life could return to its normal undisturbed state and attempt to recover. The shore stations in Spitsbergen lasted a remarkably short time. One major Dutch one on Amsterdam Island in the north-west of Spitsbergen, was established in 1619, had hundreds of ships visiting it in a season in its heyday in the late 1620s and 1630s, and yet had been abandoned before 1640 because all the whales in the coastal waters had been killed. However plundering of eggs and indiscriminate shooting of birds continued even though these stations were replaced by processing at sea. Sealers and other fishing boats continued to land when they wanted some fresh meat or even some 'sport'. Such exploitation was generally limited to the most accessible parts and those near the scene of the land-based whaling operations, so that vast tracts of the arctic fortunately remained untouched by this kind of activity.

Exploration and opening up of the arctic proceeded quite slowly and man's overall impact on the wildlife was small until the advent of aircraft. Even then the total effect was slight in the early days, though the local damage could be greater. As the number of settlements increased, so each had a certain radius round it within which the birds were disturbed if not actually killed or their nests robbed. This development has been going on at an ever increasing rate, until now there is nowhere that man cannot reach by plane or helicopter. The driving force behind this

An abandoned whaling station, the remains of an industry briefly thriving, and causing great disturbance to the wildlife of the area, before the whales were virtually exterminated by over-hunting.

rapid exploration has been the desire to discover and then exploit the mineral wealth. While this was confined to a relatively small number of mines, producing coal, lead, or a few other products, there was little threat to the wildlife, but the need for oil and the impetus that the search for it has received in the last few years has produced the greatest single danger to the whole of the arctic. The presence of oil beneath the land and sea has been known for many years but it is only recently that oil has been proved to be there in really large commercial quantities.

The early exploration for oil has left behind it ugly scars that may never heal, for the surface of the tundra is an extremely fragile and thin skin composed of an intricate web of flowering plants, mosses, and lichens, sitting on a very shallow layer of unfrozen ground overlaying the permafrost. The lichens are more numerous and wide-spread than any other plant group in the arctic and may be the dominant part of the flora over large areas. Damage an area of grass on your lawn, dig it up even, and it will recover within a year, perhaps months. Damage an area of tundra and it will take twenty or more years to recover, or may never do so. Lichens grow extremely slowly—a patch a foot across will probably have taken as many as fifty years to reach that size. The continuous blanket of the arctic tundra is thus the product of a very long process which can, however, be irrevocably damaged in moments.

The single passage of a heavy vehicle over the tundra, particularly one with caterpillar tracks, can break up the surface layer, the removal of which immediately affects the frost-bound ground below. It thaws and freezes at a different rate from the untouched tundra either side of the tracks, while the absence of plant cover permits the onset of erosion, that most insidious of nature's forces. Near some oil prospecting camps in Alaska there are six-foot-deep gulleys that only ten years ago were a pair of wheel tracks. Although the activities of vehicles, particularly tracked ones, are now proscribed in some parts, damage has already been done and more will inevitably follow.

This localised damage may not have much direct impact on birds, other than in the immediate vicinity of the prospecting or drilling camp. It is the oil itself that does so much damage. The controversy that surrounded the taking of the tanker *Manhattan* through the North West Passage round the north coast of Canada a few years ago is evidence of the alarm engendered in men's minds, and not just those of committed conservationists, at the thought of oil being shipped through arctic seas. The Alaskan oil is to be piped south, causing its own problems to wildlife on the way. But the prospecting goes on in arctic Canada and there is no guarantee that oil found there will be piped

The surface of the tundra is easily damaged by vehicles and may take decades to recover. Alternatively the tracks may deepen into permanent gullies.

south even if it is possible. Oil in cold waters does not decompose nearly as fast as it does in warmer seas, so it remains a menace for several years instead of merely months. It is horrific even to contemplate what effect a major spill within the feeding radius of one of the really large seabird colonies would have.

The creation of reserves and national parks has lagged far behind the situation in more populated lands. Perhaps this is only to be expected, but it is a state of affairs that must not be tolerated for much longer, for their need is readily demonstrable. Our state of knowledge about many birds is now good enough to be able to identify at least the principal breeding, feeding, and moulting areas. The safeguarding of these can be shown to be essential to the well-being of the particular populations. It is too much to expect that such areas will be created in localities where oil or other minerals have already been found, or that they will be sacrosanct from future exploration and development, but at least some control on activities could be exercised. Nature reserves have quite recently been declared in Spitsbergen by Norway and in Greenland by Denmark, while Canada, the United States, and Russia have long had them. But it is pertinent to ask how efficient is the policing of these reserves and how effective their status, and whether the future exploitation of mineral resources would take precedence over the resource of wildlife.

Tourism has been involved in the arctic in a small way for many years and is now on the increase, with ships visiting some areas every summer, and airstrips and hotels being opened in others. There is as yet no danger of floods of visitors, but even small numbers of people can do great damage to an Eider colony, for example by unthinkingly getting too close in order to take just one more photograph. Tourists must therefore be regulated as far as possible and channelled so that they can see all that they paid to come and see but still leave it there undisturbed for the next group.

The arctic is a wonderful place that has a habit of casting its spell over all those who go there and of creating an ill-defined but persistent desire to return. There comes a longing to experience once more the pleasures of landing on a remote shore, of seeing the pale creamy light of an arctic midnight, and of hearing sounds of myriad seabirds forming a backcloth to the nearer and sweeter song of the Snow Bunting or the trill of a Sanderling. Once on land, one walks over a carpet of mosses, lichens, and beautiful if miniature wild flowers. And one sees, not the riotous wealth of bird life that exists further south, with its dawn chorus, vivid colours, and multitude of sizes and shapes, but a smaller more tangible variety of birds, some rare, some in vast numbers, but all of which have come to terms with the exacting conditions of life there, of migration, and of seeking some far-off winter home. All are completely suited to this life and all, either individually or through their progeny, return year after year. They do it and will go on doing it for as long as we, the human race, will let them. It is an unbearable thought that civilisation should spoil this land, even though it is impossible to keep it away altogether. But at least let it come gently, and let it be controlled. Keep it within bounds, exercise great care, and there is then no reason why we and future generations of people should not enjoy the birds of the arctic and their future generations.

Bibliography

This is not an exhaustive list of references consulted but includes all the more important sources. In cases where authors have been responsible for several papers on one species or subject I have usually listed only the latest, and the interested reader will be able to find the earlier ones cited in that.

General interest, and Chapters 1 and 2

ARMSTRONG, E. A. (1954). The behaviour of birds in continuous daylight. *Ibis*, **96**, 1–30.

BAILEY, A. M. (1948). *Birds of Arctic Alaska*. Denver: Colorado Museum of Natural History.

DEMENTIEV, G. P. and GLADKOV, N. A. (1951–4). *Birds of the Soviet Union* (6 vols). Moscow: State Publishers (English version 1966–8 by Israel Program for Scientific Translations).

FREUCHEN, P. and SALOMONSEN, F. (1959). *The Arctic Year*. London: Jonathan Cape.

GODFREY, W. E. (1966). *The Birds of Canada*. Ottawa: National Museums of Canada.

IRVING, L. and KROG, J. (1956). Temperature during the development of birds in arctic nests. *Physiol. Zool.*, **29**, 195–305.

JEHL, J. R. (1970). Patterns of hatching success in subarctic birds. *Ecology*, **52**, 169–73.

JOHANSEN, H. (1956, 1958). Revision and Entschung der arktischen Vogelfauna, I, II. *Acta Arctica Fasc.*, 8, 9.

MADSEN, H. and WINGSTRAND, K. G. (1959). Some behavioural reactions and structures enabling birds to endure winter frost in arctic regions. *Vidensk. Medd. Dansk Natur. Foren.*, **120**, 15–23.

MARSHALL, A. J. (1952). Non-breeding among arctic birds. *Ibis*, **94**, 310–33.

SALOMONSEN, F. (1950). *The Birds of Greenland*. Copenhagen: Ejnar Munksgaard.

SALOMONSEN, F. (1967). *Fuglene på Grønland*. Copenhagen: Rhodos.

SALOMONSEN, F. (1972). Zoogeographical and ecological problems in arctic birds. *Proc. XVth Int. Orn. Cong.*, 1970, 25–72.

SCHOLANDER, P. F., WALTERS, V., HOCK, R. and IRVING, L. (1950). Body insulation and heat regulation of some arctic and tropical mammals and birds. *Biol. Bull.*, **99**, 225–58.

SNYDER, L. L. (1957). *Arctic Birds of Canada*. University of Toronto Press.

VIBE, C. (1967). Arctic animals in relation to climatic fluctuations. *Medd. om Grøn.*, **170**, No. 5.

WATSON, A. (1963). Birds numbers on tundra in Baffin Island. *Arctic*, **16**, 101–8.

Chapter 3

BARRY, T. W. (1962). Effect of late seasons on Atlantic Brant reproduction. *Jour. Wildl. Mgmt.*, **26**, 19–26.

BARRY, T. W. (1968). Observations on natural mortality and native use of Eider Ducks along the Beaufort Sea coast. *Canad. Fd. Nat.*, **82**, 140–4.

BOYD, H. (1966). The assessment of the effects of the weather on the breeding success of geese nesting in the arctic. *Statistician*, **16**, 171–80.

FREEMAN, M. M. R. (1970). Observations on the seasonal behavior of the Hudson Bay Eider *(Somateria mollissima sedentaria)*. *Canad. Fd. Nat.*, **84**, 143–53.

HARVEY, J. M. (1971). Factors affecting Blue Goose nesting success. *Canad. J. Zool.*, **49**, 223–34.

KRECHMAR, A. V. and LEONOVICH, V. V. (1967). (Distribution and biology of the Red-breasted Goose in the breeding season.) *Probl. Sev.*, **11**, 229–34 (in Russian).

LEFEBVRE, E. and RAVELING, D. G. (1967). Distribution of Canada Geese in winter as related to heat loss at varying environmental temperatures. *Jour. Wildl. Mgmt.*, **31**, 538–46.

LEMIEUX, L. (1959). The breeding biology of the Greater Snow Goose on Bylot Island, Northwest Territories. *Canad. Fd. Nat.*, **73**, 117–28.

MACINNES, C. (1966). Population behavior of eastern arctic Canada Geese. *Jour. Wildl. Mgmt.*, **30**, 536–53.

MACINNES, C. D. and MISRA, R. K. (1972). Predation on Canada Goose nests at McConnell River, Northwest Territories. *Jour. Wildl. Mgmt.*, **36**, 414-22.

NORDERHAUG, M., OGILVIE, M. A. and TAYLOR, R. J. F. (1965). Breeding success of geese in west Spitsbergen, 1964. *Wildfowl Trust Ann. Rep.*, **16**, 106–10.

OGILVIE, M. A. and MATTHEWS, G. (1969). Brent geese, mudflats and Man. *Wildfowl,* **20**, 119–25.

RYDER, J. P. (1969). Nesting colonies of Ross's Goose. *Auk,* **86**, 282–92.

SCOTT, P., BOYD, H. and SLADEN, W. J. L. (1955). The Wildfowl Trust's second expedition to central Iceland, 1953. *Wildfowl Trust Ann. Rep.*, **7**, 63–98.

SLADEN, W. J. L. and COCHRAN, (1969). Studies of the Whistling Swan, 1967–1968. *Trans. 34th. N. Am. Wildl. Conf.*, 42–50.

Chapter 4

BENGTSON, S-A. (1970). Breeding behaviour of the Purple Sandpiper in West Spitsbergen. *Ornis Scand.*, **1**, 17–25.

DUFFEY, E. (1950). The rodent-run distraction-behaviour of certain waders. *Ibis,* **92**, 27–33.

HILDEN, O. (1965). Zur Brutbiologies des Temminck-Strandlaufers, *Calidris temminckii* (Leisl.). *Ornis Fenn.*, **42**, 1–5.

HOHN, E. O. (1971). Observations on the breeding behaviour of Grey and Red-necked Phalaropes. *Ibis,* **113**, 335–48.

HOLMES, R. T. (1970). Differences in population density, territoriality, and food supply of Dunlin on arctic and subarctic tundra. In: *Animal Populations in Relation to their Food Resources,* ed. A. Watson. Oxford and Edinburgh: Blackwell.

HOLMES, R. T. (1972). Ecological factors influencing the breeding season schedule of Western Sandpipers *(Calidris mauri)* in subarctic Alaska. *Amer. Mid. Nat.,* **87**, 472–91.

HOLMES, R. T. and PITELKA, F. A. (1964). Breeding behavior and taxonomic relationships of the Curlew Sandpiper. *Auk,* **81**, 362–79.

HOLMES, R. T. and PITELKA, F. A. (1966). Ecology and evolution of sandpiper (Calidridinae) social systems. *Abstr. XIVth Int. Orn. Congr.*, 1966, 70–71.

JEHL, J. R. (1973). Breeding biology and systematic relationships of the Stilt Sandpiper. *Wilson Bull.*, **85**, 115–47.

KISTCHINSKI, A. A. (1975). Breeding biology and behaviour of the Grey Phalarope *Phalaropus fulicarius* in east Siberia. *Ibis,* **117**, 285–301.

MCLEAN, S. F. (1969). Ecological determinants of species diversity in arctic sandpipers near Barrow, Alaska. Ph.D. Thesis, University of California, Berkeley, Ca.

NETTLESHIP, D. N. (1973). Breeding ecology of Turnstone *Arenaria interpres* at Hazen Camp, Ellesmere Island, N.W.T. *Ibis,* **115**, 202–17.

NETTLESHIP, D. N. (1974). The breeding of the Knot *Calidris canutus* at Hazen Camp, Ellesmere Island, N.W.T. *Polarforschung,* **44**, 8–26.

PARMALEE, D. F., GREINER, D. W. and GRAUL, W. D. (1968). Summer schedule and breeding biology of the White-rumped Sandpiper in the central Canadian arctic. *Wilson Bull.*, **80**, 5–29.

PARMALEE, D. F. and PAYNE, R. B. (1973). On multiple broods and the breeding strategy of arctic Sanderlings. *Ibis,* **115**, 218–26.

PITELKA, F. A. (1959). Numbers, breeding schedule and territoriality in Pectoral Sandpipers of northern Alaska. *Condor,* **61**, 233–64.

RANER, L. (1972). Förekammer polyandri hos smalnäbbad simsnäppa *(Phalaropus lobatus)* och svartsnäppa *(Tringa erythropus)*? *Fauna och Flora,* **67**, 135–8.

SUTTON, G. M. (1967). Behaviour of the Buff-breasted Sandpiper at the nest. *Arctic,* **20**, 2–7.

Chapter 5

BATESON, P. P. G. and PLOWRIGHT, R. C. (1959). The breeding biology of the Ivory Gull in Spitsbergen. *Brit. Birds,* **52**, 105–14.

BEDARD, J. (1969). Adaptive radiation in Alcidae. *Ibis,* **111**, 189–98.

BEDARD, J. (1969). Feeding of the Least, Crested and Parakeet Auklets around St. Lawrence Island, Alaska. *Canad. J. Zool.,* **47**, 1025–50.

BEDARD, J. (1969). The nesting of the Crested, Least and Parakeet Auklets on St Lawrence Island, Alaska. *Condor,* **71**, 386–98.

DIVOKY, G. J., WATSON, G. and BARONEK, J. C. (1974). Breeding of the Black Guillemot in northern Alaska. *Condor,* **76**, 339–43.

MACDONALD, S. M. and MACPHERSON, A. H. (1962). Breeding places of the Ivory Gull in arctic Canada. *Nat. Mus. of Canada Bull.,* **183**, 111–17.

MAHER, W. J. (1974). Ecology of Pomarine, Parasitic and Long-tailed Jaegers in northern Alaska. *Cooper Orn. Soc. Pacific Coast Avifauna,* No. 37.

NETTLESHIP, D. N. (1972). Breeding success of the Common Puffin *(Fratercula arctica* L.) in different habitats at Great Island, Newfoundland. *Ecol. Mono.,* **42**, 239–68.

NORDERHAUG, M. (1970). The role of the Little Auk *Plautus alle* (L.) in arctic ecosystems. *Antarctic Ecol.,* **1**, 558–60.

SEALY, S. G. (1973). Adaptive significance of post-hatching developmental patterns and growth rates in the Alcidae. *Ornis Scand.,* **4**, 113–22.

SEALY, S. G. (1973). Breeding biology of the Horned Puffin on St Lawrence Island, Bering Sea, with zoogeographical notes on the North Pacific puffins. *Pacific Science,* **27**, 99–119.

SEALY, S. G. (1975). Influence of snow on egg-laying in Auklets. *Auk,* **92**, 528–38.

SEALY, S. G. and BEDARD, J. (1973). Breeding biology of the Parakeet Auklet *(Cyclorrhynchus psittacula)* on St. Lawrence Island, Alaska. *Astarte,* **6**, 59–68.

Chapter 6

BROOKS, W. S. (1968). Comparative adaptations of the Alaskan Redpolls to the arctic environment. *Wilson Bull.,* **80**, 253–80.

DRURY, W. H. (1961). Studies on the breeding biology of the Horned Lark, Water Pipit, Lapland Longspur and Snow Bunting on Bylot Island, N.W.T., Canada. *Bird Banding,* **32**, 1–46.

HUSSELL, D. J. T. (1972). Factors affecting clutch size in arctic passerines. *Ecol. Mono.,* **42**, 317–64.

JEHL, J. R. and HUSSELL, D. J. T. (1966). Effects of weather on reproductive success of birds at Churchill, Manitoba. *Arctic,* **19**, 185–91.

SHIELDS, M. (1969). Activity cycles of Snowy Owls at Barrow, Alaska. *Murrelet,* **50**, 14–16.

STEPHEN, W. J. D. (1967). Bionomics of the Sandhill Crane. *Canad. Wildl. Serv.,* Rep. Series No. 2.

SUTTON, G. M. and PARMALEE, D. F. (1954). Survival problems of the Water Pipit in Baffin Island. *Arctic,* **7**, 81–92.

SUTTON, G. M. and PARMALEE, D. F. (1954). Nesting of the Snow Bunting on Baffin Island. *Wilson Bull.,* **66**, 159–74.

TEMPLE, S. A. (1974). Winter food habits of Ravens on the Arctic Slope of Alaska. *Arctic,* **26**, 41–6.
WATSON, A. (1957). The behaviour, breeding and food-ecology of the Snowy Owl *Nyctea scandinavica. Ibis,* **99**, 419–62.
WILLGOHS, J. (1962). The White-tailed Eagle *Haliaetus albicilla albicilla* (Linne) in Norway. *Arbok for Universitetet i Bergen,* Serie 1961, No. 12.
WYNNE-EDWARDS, V. C. (1952). Zoology of the Baird Expedition (1950). 1. The birds observed in central and south-east Baffin Island. *Auk,* **69**, 353–91.

Chapter 7

BROOKS, J. W., BARTONEK, J. C., KLEIN, D. R., SPENCER, D. L. and THAYER, A. S. (1971). Environmental influences of oil and gas development in the arctic slope and Beaufort Sea. *U.S.D.I. Resource Publ.* 96.
HANSON, H. C. and CURRIE, C. (1957). The kill of wild geese by the natives of the Hudson-James Bay region. *Arctic,* **10**, 211–29.
KLEIN, D. R. (1966). Waterfowl in the economy of the Eskimos on the Yukon-Kuskokwim Delta. *Arctic,* **19**, 319–36.
Proceedings Conf. Productivity and Conservation, Northern Circumpolar Lands, Edmonton 1969. *I.U.C.N. Publ.* 16.

Picture acknowledgements

Page numbers given; those in italics refer to colour.

T. Andrewartha: *26–7* (bottom), *37*, *178*.
Ardea Photographics, London: 195; Dennis Avon & Tony Tilford 41; J. A. Bailey *104*; J. B. & S. Bottomley 47, 88, 120, 124, 137, 141; Dr Kevin J. V. Carlson A.R.P.S. *55* (bottom), 174; M. D. England 192, 201; K. W. Fink *25*, 36, *73*, *149*; Gary R. Jones 35; C. R. Knights *208*; Peter Lamb *132*; Ake Lindau 123, *177*; John Marchington 2–3, 71; T. Marshall *114*, *189*; E. Mickleburgh 214; S. Roberts *55* (top), *178–9*; B. L. Sage 15, 17, 19, 21 (both pictures), 22, 34, *101*, 184, 215; R. Vaughan 87, 138, 153; Tom Willock *180*; John Wightman 157.
J. B. & S. Bottomley: 33, 93, 99, 105, 115, 116, 127 (top), 139, 145, 192–3, 193, 196, 203.
Bruce Coleman Limited, Uxbridge: Jen & Des Bartlett *113*; David & Katie Urry *150*.
L. R. Dawson: 62.
B. Gates: 31.
Pamela Harrison F.R.P.S.: 52, 54, 76, 78, 79, 81, 82, *91*, 117, 154.
Morley Hedley F.R.P.S.: 29, 60, 67.
Eric Hosking F.R.P.S.: *56*, 75, 76–7, 136, 142, 144, 147, 148, 160, 163, 164 (top), 169, 172, 173, 175, 181, 183, 191, 198, *207*.
E. E. Jackson, Wildfowl Trust: *38*.
Naturfotograferna, Österbybruk, Sweden: Erik Isakson 32.
M. Norderhaug: 158.
M. A. Ogilvie: *26–7* (top), *28*, *102–3*.
Photo Researchers Inc., N.Y.C.: Ken Brate *190*; G. C. Kelly *168*.
Niall Rankin, photographs supplied by Eric Hosking: 44, 65, 86, 152.
C. P. Rose: 212.
B. L. Sage: 49, 59, 64, *74*, 85, 108, 118, 121, 127 (bottom), *131*, 164 (bottom), *167*, 185.
W. Stribling, photograph supplied by B. L. Sage: 95.
Wildfowl Trust, Slimbridge, Gloucestershire: 16, *92*.

Maps and endpapers by Neil McConachie
Other line drawings by Stephen Chapman

Index

Page numbers in bold show where species are described in detail; those in italic indicate pictures or captions.

Index

Index